A Futurist's Guide to Emergency Management

A Futurist's Guide to Emergency Management

Adam S. Crowe

CRC Press
Taylor & Francis Group
Boca Raton London New York

CRC Press is an imprint of the
Taylor & Francis Group, an **informa** business

CRC Press
Taylor & Francis Group
6000 Broken Sound Parkway NW, Suite 300
Boca Raton, FL 33487-2742

Printed on acid-free paper
Version Date: 20150102

International Standard Book Number-13: 978-1-4822-5391-7 (Hardback)

Library of Congress Cataloging-in-Publication Data

Crowe, Adam S.
A futurist's guide to emergency management / Adam S. Crowe.
pages cm
Includes bibliographical references and index.
ISBN 978-1-4822-5391-7 (alk. paper)
1.Emergency management--Planning. 2.Emergency management--Forecasting. I. Title.

HV551.2.C7624 2015
363.34'8--dc23 2014048257

Visit the Taylor & Francis Web site at
http://www.taylorandfrancis.com

and the CRC Press Web site at
http://www.crcpress.com

CONTENTS

INTRODUCTION

> As always, new technologies hold the promise of doing great good, supplying new sources of clean energy, curing disease, and otherwise enhancing our lives. From experience, however, we also know that new technologies can be used to diminish humanity and destroy societies. We can manage our technology or become victims of it. The choice is ours, and the Clock is ticking.
>
> ~Science and Security Board of the Bulletin of Atomic Scientists [1]

WHY THE FUTURE IS IMPORTANT NOW

The future is impossible to know, particularly when it comes to the dynamic situations faced by communities as they prepare to respond to the various dynamic events that can disrupt the local quality of life. However, that does not mean the future can or should be ignored. Community leaders and their designated emergency managers must use subjective and objective analyses to trend, forecast, predict, and project the various conditions that directly or indirectly create risk and increased vulnerability within a given community. These conditions include the various psychological, physical, economic, social, and sociodemographic characteristics of the people, environment, and culture of every community.

Interestingly, the foundation of these futuristic projections is in the present. Numerous studies, anecdotal observations, and various lessons over the past decades have set forth patterns and collections of information that have begun to build trends toward the next few years or even decades from now. Much like a threat assessment, which helps provide focus to planning and resource priorities in the future, evaluating future trends can help provide these same types of clarity and perspective to emergency managers at all levels.

Unfortunately, when most people think about the future they think about far-flung and outlandish developments out of science fiction stories like *Star Trek* or *The Jetsons*. Certainly there are stories in the news every day about seemingly impossible items like a team of scientists from Japan's Osaka University developing technologies to attach fuel cells to the backs of roaches to create so-called "cybugs" [2]. Conversely, others may envision a future similar to the dystopian societies (and often disasters) in movies like *Mad Max, The Matrix*, or *The Day After Tomorrow*. For example, in 2007, former vice president and environmental "guru" Al Gore predicted that the North Pole would be ice free by 2013, with sea levels rising by 20 feet (which ultimately did not come to pass) [3]. Obviously those "futures" are possible, but are solely based on conjecture and wild guesses.

Clearly, there are challenges to projecting and forecasting not only the events of the future, but also how they may impact society at large. However, specific strategies were applied throughout this book to minimize these challenges as much as possible. These strategies include a focus on emerging technologies, consideration of technological applications, and identification of all market forces, by focusing on realistic rather than imaginative directions. This systematic approach was applied across all three sections of this book, whose themes are citizens, technology, and the future (Section 1), preparedness, response, and recovery (Section 2), and emerging global threats (Section 3).

UNDERSTANDING FORECASTING AND PREDICTION

This futurist guide, as well as all assessments of the future, must be based on as much science as possible. Without grounding future direction in analysis that is as objective as possible, there is a significant risk that subjective views and superstition can make any assessment no better than the science fiction and fantasy that fill books and movie theaters. This distinction is a fine line when looking toward the future, but is most effectively delineated by utilizing tools such as current statistics, predictive modeling and forecasting, process analysis, and organizational intelligence.

The simplest of these approaches is the use of statistics and patterns of current activities. This will serve as the foundational evidence for all of the future trends that will be considered in this book. These statistics will be based on well-crafted empirical research and anecdotal behaviors that have widespread acceptance or defendable stances within academic, research, and practical programming. As the breadth of these foundational statistics widens, the possible futurist projects also increase, which helps create clarity in a projected and interconnected future.

Once the statistics for a given issue are established, predictive modeling and forecasting must be initiated to begin a reasonable and fair approach to futuristic projections. Predictive modeling typically identifies underlying relationships in statistics and historical data that can then be mathematically represented. This mathematical representation can then be utilized for forecasting or classification for future events [4]. The most commonly recognized form of predictive modeling is related to day-to-day and severe weather patterns that are produced by the National Weather Service, media meterologists, and commercial weather forecasting companies. These groups use scientific observations, tests, and data to project future activities and trends. Without fail, the accuracy of these predictions decreases the farther out they are projected. One National Weather Service official was quoted in a 2013 *Washington Post* article saying, "We sustain higher accuracy out to two to three days in advance; then it starts dropping off faster at days six through eight" [5]. Conversely, the National Weather Service recognizes that specific forecasts can only happen in the short term while more extreme events such as hurricanes receive longer term predictions that are often extremely vague.

Predictive modeling is frequently utilized for mission-critical operational decisions to help prioritize decision making in both near-time and long-term planning efforts. This application is often used by emergency managers and homeland security officials in day-to-day intelligence and operational decisions like the weather forecasting mentioned earlier, as well as in long-term planning and resource allocations related to community-based threat and risk assessments. This type of predictive modeling will also be highly valuable as emergency managers attempt to address the trending and future challenges that are discussed in this book. From topics like the rise of smart devices (Chapter 1), predictive behavior (Chapter 3), the "Internet of Things" (Chapter 4), "Black Swan" events (Chapter 9), climate change (Chapter 11), and cyberthreats (Chapter 13), these predictive models will be used as often as possible to project planning and resource needs for professional emergency managers.

The next component of forecasting is predictive analytics and optimization. Instead of trying to forecast or predict how technology or environmental cues change in the near and far future, this concept helps identify patterns in the ways or methods that such technologies are utilized. These patterns are often predicted through complex mathematical systems or equations called algorithms. These types of algorithms are commonly utilized by commercial products like Netflix or Match.com to predict viewer choices or likely romantic matches [4]. They leverage past choices and the choices of others to forecast or predict future choices. These types of programs are critical to customer service for these companies to improve the success and positive engagement for the client base.

The challenge to predictive analysis in its most common form is that it is dependent on highly technical math processes, which are unavoidable for most government and public safety agencies. However, there is a growing class of technologies that can be leveraged to help these smaller organizations apply it for use to predict the behavior of citizens and constituents before, during, and after a disaster. This phenomenon is discussed in Chapters 2 (communications) and 3 (data mining and predictive behavior).

The last two types of predictive analysis and forecasting are analysis and organizational intelligence. These are the most abstract features and thus are farthest from the raw data, but potentially most applicable to organizations seeking to identify issues in the future. One approach to analysis is the use of data or behavioral modeling. One of the best examples of predictive modeling is the Google search engine. Over the years, Google has shifted its modeling and predictions to ultimately create a search engine that "understands exactly what you mean and gives you back exactly what you want" [5]. Although no emergency management or public safety agency has the processing or modeling power of Google, it is an ideal (perhaps utopian) goal to give disaster survivors exactly what they need when they need it. This is particularly important given the limited resources available before, during, and after a disaster. This type of modeling is of particular importance in Chapters 8, 9, 10, and 11 as the impact of recovery, perception of risk, disaster economies, and so-called Black Swan events are considered.

By utilizing forecasting, predictive analytics, and modeling, an organization of any type or size increases its organizational intelligence. The concept of organizational intelligence is focused on the ability to identify and cultivate knowledge and apply it in strategic and targeted ways to meet organizational goals. The concept of organizational intelligence is a shift from traditional organizational models (and most professional emergency management and homeland security programs), which were often viewed simply as a collection of people and resources applied to tasks and products. The need for organizational intelligence is particularly important during emergency and disaster preparedness, which involves the interaction of a multitude of individuals, systems, and organizations within a local or broader organization [6].

Leveraging these prediction tools and understanding the future is the focus of this book. It is critical for emergency management and homeland security managers and organizations to understand these tools so they can more accurately look to the future and begin to more definitively predict the issues and community challenges that will present themselves in the near future. Each chapter considers the current trends, applies various models, and identifies ways to improve the reader's personal and organizational intelligence.

RELEVANCE TO DISASTER MANAGEMENT

Emergency management and disaster response professionals do not often take the time to look to the future to understand the impact on personal and/or organizational success. In many ways this is a fundamental flaw of an industry that is often burdened by limited personnel, resources, equipment, and political sway to move beyond the most imminent and pressing threat or issue. While a realistic challenge, it often leads to short-sighted decisions about planning, preparedness, response, and risk reduction in a given community. This issue is commonplace throughout the history of managing the impacts of emergencies and disasters.

Starting with the rise of the Cold War, the profession of emergency management almost solely focused on the threat of nuclear war. This threat was addressed through comprehensive civil defense programming and planning. As the threat of nuclear war diminished in the 1970s and early 1980s, there was an industry shift to encourage a broader range of issues (mostly natural hazards) that most commonly was called emergency preparedness. Unfortunately, this shift was not based on sound analysis or research. While there were a few noteworthy social science researchers, like E. L. Quarantelli, Russell R. Dynes, and Gilbert White, the practical application and acceptance by emergency managers lagged by many years (and some would argue continue to lag even today) [7].

This type of divergence between academic research and practical application was repeated in the early 2000s after the September 11 terrorist attacks. Specifically, many organizations shifted philosophical models away from all-hazards or natural disaster models of planning toward homeland security and the prevention of terrorism. After a decade of this model and a massive reduction in homeland security funding, many emergency managers have been left in the lurch with many still lacking a natural and practical connection to academic studies and no financial incentives to direct philosophical approaches. This dynamic creates a phenomenal opportunity to look to the future and evaluate where the profession is going and what choices can be made to adequately address the circumstances and issues that rise from it.

The need for disaster managers to look to the future is critical to the practice of professional emergency management in countries throughout the world. Countries including the United States, Canada, Australia, New Zealand, and other areas in Asia and Europe have strong or developing emergency management programs. While there are some differences in the application of modern emergency management principles in these areas, there are two facts on which professional emergency managers and researchers can agree: (1) There are fewer deaths from disasters and (2) the costs of disasters are increasing. For example, Swiss Re (the world's second largest reinsurer) estimated that economic losses from global disasters in 2013 reached $130 billion, of which only 34% was covered by insurance [8]. According to the International Disaster Database from the Center for the Research on the Epidemiology of Disasters (CRED), the number of reported disasters has exponentially risen from a negligible number at the turn of the twentieth century to more than 350 events per year by 2011 with the growth curve started after World War II [8]. Likewise, CRED found that the total global cost of disasters rose exponentially in the 1940s to a figure exceeding $100 billion by 2011. Moreover, during that same period of review, CRED discovered that the number of people killed from global disasters fell from close to 500,000 in 1900 to only several thousand in 2011 [9].

In addition to these trends, the rise of social media (and related technologies) has had a profound impact on the management of emergencies and disasters throughout the world. In many ways, the communication and operational impacts of social media on all phases of emergency management have been revolutionary and are still not fully understood or accepted by all professional organizations. According to some professionals, this impact is on par with previous communication revolutions like the printing press, radio, television, and cable news [10]. These impacts have left the industry slightly off balance as it has sought ways to understand and apply social media principles in step with its growth and application. This concept as well as a variety of other digital considerations is expanded upon in Chapter 1.

One characteristic that has driven and will continue to drive some of the challenges related to adoption of social media and other technologies is the changing social and demographic factors within domestic and international communities. Specifically, additional generations are included in workforces and the community at large as younger citizens adopt technologies earlier and older citizens stay engaged in community leadership and decision making longer as health issues and life expectancy continue to improve. These sociodemographic issues are further compounded by changing cultural standards, which ultimately impact disaster preparedness and identification of needs before, during, and after an emergency or disaster. These issues are further expanded upon in Chapter 5.

This book is about why the future is important today. Inherently, some of the modeling and projection ultimately presented within the chapters will be wrong. In truth, they may even be laughable upon reflection 10, 20, or 50 years from now. At the same time, the forecasting utilized in this book is based on the best available data and knowledge about the dynamics of the evaluated event. Consequently, like all good preparedness strategies, the efforts of considering the potential impacts and adjusting resources and planning strategies to meet those impacts will always improve an organization's readiness to respond to any emergency or disaster.

REFERENCES

1. Pappas, Stephanie. (2014). "Doomsday Clock Set at 5 'Til Midnight." Yahoo News. http://news.yahoo.com/doomsday-clock-set-5-39-til-midnight-231118389.html. Accessed February 28, 2014.
2. Heater, Brian. (2014). "Fuel Cells Powered by Roach Blood May Someday Lead to Useful Networks of Cyborgs." Yahoo Tech News. https://www.yahoo.com/tech/fuel-cells-powered-by-roach-blood-may-someday-lead-to-75485513322.html. Accessed March 1, 2014.
3. Editorial: "Al Gore, Soothsayer." (2013). The *Washington Times.* http://www.washingtontimes.com/news/2013/dec/16/editorial-al-gore-soothsayer/. Accessed March 1, 2014.
4. Kalakota, Ravi. (2002). "Predictive Analytics 101." Business Analytics 3.0 Blog. http://practicalanalytics.wordpress.com/predictive-analytics-101/. Accessed March 3, 2014.
5. Palmer, Brian. (2013). "Long-Term Weather Forecasts Are a Long Way from Accurate." *Washington Post.* http://www.washingtonpost.com/national/health-science/long-term-weather-forecasts-are-a-long-way-from-accurate/2013/04/15/1f9a2ac8-a05b-11e2-be47-b44febada3a8_story.html. Accessed March 6, 2014.
6. Halal, William E. (1997). "Organizational Intelligence: What Is It, and How Managers Can Use It." Strategy+Business. http://www.strategy-business.com/article/12644?gko=4a546. Accessed March 11, 2014.
7. Bandel, Carolyn. (2014). "Swiss Re's Cole Says Uninsured Disaster Losses Growing." *Bloomberg News.* http://www.bloomberg.com/news/2014-01-15/swiss-re-s-cole-says-uninsured-disaster-losses-growing.html. Accessed March 12, 2014.

8. Oyola-Yemaiel, Arthur and Wilson, Jennifer. (n.d). "Social Science Hazard/Disaster Research: Its Legacy for Emergency Management Higher Education. *The Future of Emergency Management*. North Dakota State University. https://www.google.com/url?sa=t&rct=j&q=&esrc=s&source=web&cd=1&cad=rja&ved=0CCsQFjAA&url=https%3A%2F%2Fwww2.ed.gov%2Frschstat%2Fresearch%2Fpubs%2Fguide-emergency-management-pwd.doc&ei=UOcTU6i4MaL70gGWiICoDw&usg=AFQjCNF49sIZaM9j1UKBEIRezQwyoTltsw&sig2=40YzYBvJK_AP-UHP1NLq7A&bvm=bv.61965928,d.dmQ. Accessed March 12, 2014.
9. "Natural Disaster Trends." (2011). International Disaster Database from the Center for the Research on the Epidemiology of Disasters. http://www.emdat.be/natural-disasters-trends. Accessed March 4, 2014.
10. Crowe, Adam. (2012). *Disasters 2.0: The Application of Social Media in Modern Emergency Management*. Boca Raton, FL: CRC Press.

THE AUTHOR

Adam Crowe is a certified emergency manager (CEM), associate business continuity professional (ABCP), and master exercise practitioner (MEP). He is a nationally recognized leader in social media and emergency management and routinely engages in cutting-edge planning, training, and exercise opportunities that improve the readiness of communities, organizations, and businesses to respond to and recover from emergencies and disasters.

He has spoken at over 50 regional, statewide, or national conferences on how social media impacts emergency management and has been published more than a dozen times in professional journals. In addition to *A Futurist's Guide to Emergency Management,* he is also the author of *Disasters 2.0: The Application of Social Media in Modern Emergency Management (2012)* and *Leadership in the Open: A New Paradigm in Emergency Management (2013).*

He currently serves as the director of emergency preparedness at Virginia Commonwealth University and lives with his wife and children in Richmond, Virginia.

Section I

Citizens, Technology, and the Future

1

The Super Digital Age

> We are constantly adding new gadgets. Even as they have gotten simpler individually, the cumulative complexity of all of them is increasing.
>
> ~Clive Thompson [1]

HISTORY OF TECHNOLOGY IN DISASTER MANAGEMENT

Since the earliest foundations of professional emergency and disaster management in the 1950s, technology has played a role in preparedness, response, and recovery from disasters as well as the reduction of risk in a given community. Some of these technologies have long been replaced, while others have remained with minimal change of approach or capabilities. However, with the exponential rise in digital technologies and the institution of social media systems since the turn of the twenty-first century, technological systems and public expectations related to those systems have drastically changed.

One of the most traditional of these technologies is the outdoor warning siren. Because of the limited other notification systems available, outdoor warning sirens ultimately became one of the symbols of the Cold War. This historical connection primarily rose from the passage of President Harry Truman's Civil Defense Act in 1950 (see Figure 1.1), which called for the establishment of large outdoor sirens [2]. The first deployed outdoor siren was utilized in Detroit, Michigan, and was a Chrysler product [2]. While the sirens were initially utilized solely to warn communities of

Figure 1.1 The national use of outdoor warning sirens was first established by President Harry S. Truman when he signed the Civil Defense Act in 1950. (Source: National Archives and Records Administration.)

nuclear threats, by the 1970s the purpose was expanded to include severe weather events and tornadoes [2]. However, by the end of the Cold War in the 1980s, these sirens became solely utilized for tornadoes and other imminent threats to life and safety.

Interestingly, these civil defense sirens not only served as the primary historical public notification system for a variety of communities, but they also became the visual standard of emergency preparedness and later the sign of antique strategies for public notification and warning. This is particularly true in light of the social media and digital technologies that will be evaluated in this chapter. This juxtaposition is no more evident than in the city of Los Angeles, which is littered with hundreds of outdoor warning sirens that have long been disconnected due to repeated reliability issues and measured deterioration even though the outdoor warning siren system was state of the art in production and capability when it was installed [3].

From an emergency management perspective, outdoor warning sirens served as the primary, but not the only communication systems. For example, printed materials, television, and radio represent the three foundational elements that have always been utilized to communicate issues to a public constituency. These types of communication systems were solely utilized until the advent of the Internet and its widespread application in the 1990s and early 2000s. At that point in time, many preparedness organizations adopted websites and other Internet-based protocols for the distribution of information. While the growth in the Internet and related browser-based information was in its own way revolutionary, it still ultimately followed the same model of distribution as the foundational systems. Primarily, information was only available to be disseminated or pushed to an intended audience or community. This one-way limitation was addressed with the astronomical growth of social media systems like Facebook and Twitter that allowed for a give-and-take, two-way communication system for anyone to use. These social media systems did not change the pre-existing technology available for communications and warning, but rather changed the dynamics of how communities expected to receive and engage in disaster-related information [4].

SOCIAL MEDIA, TEXTING, AND SMARTPHONES

Likewise, in 1984, a full two decades before the rise of social media systems, a Franco-German global mobile communications company developed a short message system that ultimately became known as texting. It took nearly 6 more years before an actual text message (specifically, "Merry Christmas") was sent between two telecommunication officials. It was not until 1993 that a handset manufacturer actually developed a cell phone device capable of sending and receiving text messages, as most of the initial network technology was merely to support the notification of voice mail messages. By 1997, Nokia released the first device with a full keyboard that was able not only to send text messages, but also to encourage their creation and distribution [5].

Public use of text messaging was slow in developing even after the devices contained the technological capability. For example, in 1995 the average American user sent 0.4 texts per month. By 2000, the average number of text messages increased to 35 per month per user [5]. This slow progression continued into 2006 where Americans sent and received approximately 65 messages per month. However, over the next 2 years there was an exponential growth in the use of text messages. Specifically, by the second quarter of 2008, American mobile subscribers sent and received an average of 357 text messages per month. This figure is even more impactful when considering that those same mobile phone users only made or received 204 phone calls per month—clearly indicating a paradigm shift in how people communicate and how they use their mobile phones [6] (see Figure 1.2).

This trend does not appear to be slowing down. According to the CTIA (the wireless industry trade association) American cell phone subscribers send more than 75 billion text messages per month and average approximately 2.5 billion messages per day [5]. Likewise, of the 90% of all Americans who owned cell phones in 2014, nearly 81% of those owners utilized text messaging on a regular basis [7]. Additionally, this trend does not appear to be slowing down as younger generations maintain cell phones and utilize text messaging at a higher rate than older generations.

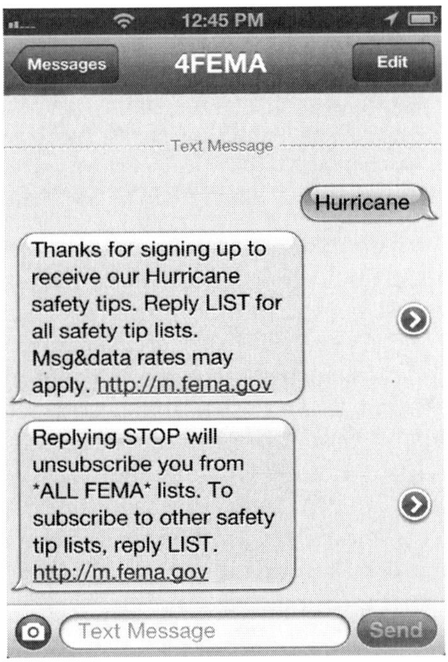

Figure 1.2 FEMA and other emergency management agencies have begun to utilize text messages to distribute information before, during, and after events. (Source: FEMA.)

For example, an average American teenager sends nearly 2,000 text messages per month [5]. This high-level adoption has significantly changed how the public expects to receive information not only from friends and family, but also from government officials before, during, and after emergencies and disasters.

Not surprisingly, the exponential rise in the use of text messaging coincides with the establishment of social media as a primary and ubiquitous tool of modern communication. As a form of two-way communication that exceeds the capabilities of broadcast systems available through traditional media forms (e.g., television and radio), social media systems have radically changed how people send and receive information on a day-to-day basis. Earlier social media systems like Friendster and MySpace started in the early 2000s and quickly grew in popularity and usage with Google attempting to buy Friendster for $30 million [8]. Unfortunately, these particular systems faded before the decade was over as more flexible and dynamic systems like Facebook and Twitter grew.

With the slight delay in development, Twitter and Facebook quickly and firmly replaced the earlier systems as the primary social media systems for the United States and then the majority of the world (see Figure 1.3). For example, according to Facebook there were 945 million monthly mobile users, 757 million daily active users, and 1.23 billion monthly active users by the end of 2013 with approximately 81% of all users

Figure 1.3 Social media systems like Facebook and Twitter have become increasingly valuable before, during, and after emergencies and disasters. (Source: FEMA/Patsy Lynch.)

located outside North America [9]. Likewise, Twitter maintains more than 645 million registered users who send 58 million tweets per day on average of which 43% originate from mobile devices [10]. These statistics are particularly impressive considering Facebook reached one million users at the end of 2004 and one billion users by September 2012 [11]. Similarly, Twitter grew from 30 million to 68 million to 138 million to 204 million active users between 2010 and 2013 [12].

In the case of Facebook, Twitter, and short-message texting, the peak in usage and users seen toward the end of the first decade of the 2000s was in conjunction with the rise in the availability and use of mobile devices. Mobile phones have come a long way since their invention over 40 years ago. The first cell phones weighed approximately 2.5 pounds and cost nearly $4,000 ($9,000 with inflation), while the most recent Apple iPhone weighed less than 4 ounces (about the weight of two eggs) and was provided for free by many mobile phone service providers [13]. While the physical changes are impressive, mobile devices have also increased in prevalence in day-to-day life. According to the Pew Research Internet Project, over 90% of Americans owned a cell phone with more than 62% of those devices being "smart" [14]. These smartphones (or other similar devices) can operate interactively and autonomously with other devices or networks through technological protocols like Bluetooth, NFC (near field communication), Wi-Fi, and LTE (long-term evolution). This connectivity allows for a push and pull of information via e-mail, social media, text messaging, operational applications, and other browser-based information. While that number dips slightly for older generations and some rural areas, it remains high in all sociodemographic categories.

Availability of these systems is merely a component of the impact of mobile devices. Because of the high portability of these devices, the widespread capability to access the Internet (or web-based sources), and simultaneous rise in communication systems like text messaging and social media, Americans use their smart devices for a variety of productivity functions including messages, Internet, using applications, listening to music, associating themselves geographically, using maps, and otherwise searching for more mundane information like weather and news. For example, according to Pew Internet, 74% of adult smartphone owners over the age of 18 regularly use their mobile phones to get directions or other information (e.g., locating restaurants) based on their current geographic location [14]. Additionally, 80% of mobile phone owners reported that they utilized their devices while watching television while another 40% reported sleeping with their phones near them to ensure they did not miss a notification [15]. While not universal, the presence of these devices and their influence on an average person are extremely impactful with the presence of mobile phones nearly as ubiquitous as other forms of communication like television, radio, and print media.

Given the widespread use and availability, mobile phones have served to unyoke the social media and traditional communication systems from the stationary desktop computers or semiportable laptop computers to which they were previously limited. In fact, Twitter and Facebook are accessed via mobile devices by their users 60% and 78% of the time, respectively [15,16].

In Other Words...Impact of Smartphones

Over the past few years, one of the most important shifts in the digital world has been the move from the wide-open Web to semi-closed platforms that use the Internet for transport [of information] but not the browser for display. It's driven primarily by the rise of the iPhone model of mobile computing and it's a world where Google can't crawl, one where HTML [code] doesn't rule. And it's the world that consumers are increasingly choosing, not because they're rejecting the idea of the Web but because these [mobile] dedicated platforms often just work better or fit better into their lives (the screen comes to them, they don't have to go to the screen). The fact that it is easier for companies to make money on these platforms only cements the trend.

~Chris Anderson, *Wired* Cofounder [17]

ELIMINATING THE DIGITAL DIVIDE

The changes in technology have also helped create a clear and definitive bridge across the so-called "digital divide." The concept of a digital divide was first identified by the *New York Times* in January 1996 in an article called "A New Gulf in American Education, the Digital Divide," which compared the availability of computers and Internet access at two California primary schools [18]. Specifically, the article and later the broader concept of a digital divide looked at social, economic, and demographic considerations that limited certain individuals from having access to Internet-based information. To put this in perspective, the original *New York Times* article quoted Reverend Jesse Jackson and the NAACP's Kweisi Mfume, who called the digital divide "classic apartheid" and "technological segregation," respectively [18] (see Figure 1.4). Likewise, another report from the state of Georgia entitled "A Nation Ponders Its Growing Digital Divide" reported that only 9% of American classrooms had access to the Internet [18].

The public and political attention given to the digital divide shifted significantly by the turn of the century as the concept fell out of political favor. This shift coincided with increased political support for marketplace solutions and a general acceptance of any digital divide being part of the "American Way" [19]. As the political outlook and philosophy changed, funding and support for programs founded in the late 1990s by the federal government like the National Telecommunications and Information Administration (NTIA)'s Technology Opportunity Program and HUD's Neighborhood Networks Program were suspended or allowed to fade away due to lack of funding [18,19]. This change was so definitive that the phrase "digital divide" was replaced by "digital opportunity" by the federal government in an attempt to frame the challenge and present a "blandly positive spin on all things computer related" [19].

Like many of the issues discussed throughout this book, the presence and availability of technology do not exist in a vacuum. There is little argument that since the Internet became available to the general public there is a divide between those who utilize technology and those who do not. Unfortunately, it is not necessarily limited to the sociodemographic

Figure 1.4 Civil rights leaders like the NAACP's Kweisi Mfume have referred to the technological challenge of the digital divide as "technological segregation." (Source: US National Oceanic and Atmospheric Administration [NOAA].)

issues that politicians would like to present. For example, a digital life columnist for the *Seattle Times* named Monica Guzman suggests that the digital divide is actually made up of four types of divisions based on the technological access, willingness to exchange information, digital identity, and technological creation [20].

Therefore, the first and most straightforward component of the digital divide is simply the availability of technology to the general population. As established earlier in this chapter, the presence of mobile phones in America is nearly ubiquitous with only minimal decreases based on gender, ethnicity, or geographic area (see Figure 1.5). This trend is no different when access to devices is looked at globally. For example, one study indicates that there are currently 4.3 billion people worldwide using mobile devices with that number rising to 5.1 billion by 2017 [21]. Unfortunately, the mere access to mobile devices is an inferior analysis as Guzman points out, "When it comes to prosperity under technology access is not the finish line, but [rather] the starting point" [20].

For example, a University of Washington professor named Ricardo Gomez has identified a growing trend he calls "pushback," which represents a growing tendency of some individuals with access to technology that intentionally resists or reduces their own access [21]. This pushback is not related to technological frustration or high costs as one might predict, but rather is predominantly due to emotional dissatisfaction with the needs being met by the technology. Likewise, Gomez found that other

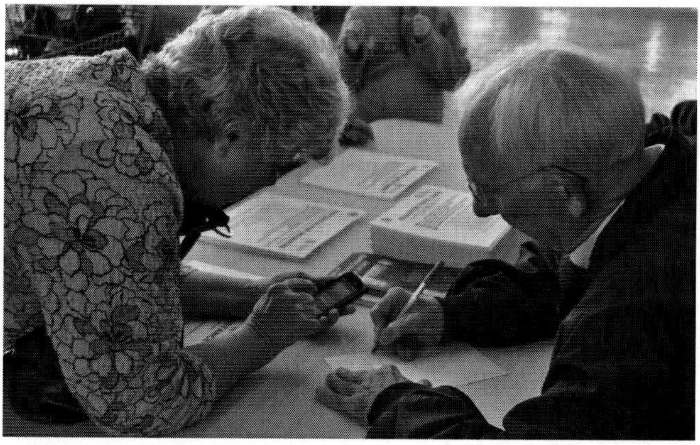

Figure 1.5 The use of cell phones differs based on age, gender, and other socioeconomic factors. (Source: FEMA/Sharon Karr.)

reasons for this pushback were related to political, religious, or moral concerns, but not necessarily related to concerns about privacy [21].

Interestingly, the US federal government has also made a significant push at eliminating economic barriers to the ownership and availability of mobile devices. Specifically, since 2005, the US Federal Communications Commission (FCC) has applied an adjusted definition of a 20-year-old "Lifeline" program that allowed qualified low-income (no higher than 135% of federal poverty guidelines) consumers access to telephone equipment and service to support finding a job, connecting with family, and accessing emergency services [22]. By 2014, nearly 92% of all qualifying households utilized the Lifeline program to acquire a phone service [22]. Specifically, this program supports devices and services including phone calls, text messages, and data exchange, which support the full integration of mobile devices into all economic strata [23].

In addition to the availability of technology, an additional component of digital divide is the willingness to exchange information. Given the widespread use of social media and mobile devices, willingness to share information is a critical element to the future of these devices and the continued development or eventual erosion of the super digital age that currently intertwines life and society. Because of the interconnectedness of social systems like Facebook and Twitter, there is inherently an openness of information when these systems are used actively. For example, Facebook's proclaimed mission is to "give people the power to share and make the world more open and connected" [9]. This openness is intrinsically available as an exchange for personal, family, and systematic privacy. This balance of openness and privacy is not just limited to engagement in social media systems, but also often pervades third-party commerce systems like online retailers who collect data and interact with pre-existing social media networks of their customers and clients.

In Other Words...The Impact on Personal Privacy

Our unbridled love affair with all things technological has an evil twin: a seemingly unstoppable encroachment on our personal privacy. The same streaming video technology that allows grandma and grandpa to chat with their grandchildren is being used to spy on employees in the workplace or capture unsuspecting lovers stealing a kiss.

~MSNBC Op-Ed, December 2000 [24]

To put this level of sharing in perspective, more than 60 hours of videos are uploaded to YouTube every minute, which is approximately more video in 1 month than the ABC, CBS, and NBC television networks created in their first 60 years combined. Likewise, 500 years of YouTube videos are watched every day on Facebook and over 700 YouTube videos are shared on Twitter each minute [25]. This level of information sharing is also prevalent on Facebook. For example, at the beginning of 2014 the average Facebook user maintained 130 friends, and 80 connected events. Moreover, there are more than one million links, two million friend requests, and three million internal messages sent on Facebook every 20 minutes [26].

The final component of the individual's willingness to share is looking at not only the quantity, but also the type of information shared. Sharing on social media systems can include a variety of documentary information (e.g., photos), but also very specific and personal types of information. This shift toward less privacy and more openness is no more evident than looking at how teenagers are utilizing these systems. For example, Pew Internet found that 71% of teenagers post their school name and city or town where they live, 53% post their personal e-mail address, and 20% post their cell phone number, which are all significant increases over the last 5 years. Additionally, 82% of teenagers posted their real birth date and name, 62% reported their relationship status, 24% posted personal videos of themselves, and 16% automatically post geolocations for individual messages or activities [27].

Interestingly, this collective shift toward increasing openness and corresponding transparency is not limited to information about an individual and/or his or her friends. Specifically, there is a growing desire for ad hoc transparency—particularly in government—for all types of information including financial, governance, and public safety data. When this expectation is not fully addressed, it puts secured information and data at risk. For example, former National Security Agency (NSA) analyst Edward Snowden and Private Bradley Manning (see Figure 1.6) both illegally released classified files from various government agencies including the NSA. While Manning was convicted of crimes under the Espionage Act and Snowden has been granted international asylum from extradition to the United States, both of these individuals have received significant support from the broader online and social system community. In particular, Snowden's philosophy is the most enlightening to understand the broader philosophical shifts in regard to the prevalence of share information, its impact on those engaged in an open community, and what

Figure 1.6 Some individuals, like former US Private Bradley Manning, have illegally released classified files from various governmental agencies. (Source: US Army.)

responses have begun to be normalized within society. Specifically, Snowden claimed during a series of interviews, "I'm willing to sacrifice [my former life] because I can't in good conscience allow the US government to destroy privacy, Internet freedom, and basic liberties for people around the world with this massive surveillance machine" [28].

In Other Words...Privacy in the Big Data Era

It isn't just privacy that is at risk in this new era of Big Data collection. Secrecy is a casualty too. It used to be classified documents were kept in a safe and seen by a select view. Now a top secret document can be accessed by hundreds, if not thousands, all with the click of a mouse.

~NPR's All Things Considered, June 2013 [29]

DIGITAL IDENTIFICATION

The third leg of the digital divide is the growth and development of a digital identification that is unique and special to each individual user. Identity is the collection of characteristics which are inherent to a respective component or intentionally or arbitrarily assigned by collective groups [30]. In the real word, names, gender, race, and ethnicity are examples of common identification strategies applied to and understood by large groups of people. This collection of characteristics impacts interactions due to natural and inherent associations which are created during exchanges—whether organic or commercial. However, digital information and related digital transactions lack those physical characteristics and associations as fundamentally the data are simply bundled packets of information [30].

That is not to say that social media users do not maintain systematic identification in digital environments. In the early days of social media, this identification was limited to the individual systems that created personal information (e.g., name and passwords) to allow specific and targeted use of that particular system. While this approach is still widely utilized as new or modified systems are added to the multitude of social media and digital systems, there are the beginnings of a unification of these system identifications. Specifically, Facebook's Open Graph system allows for a personalization that is socially based, but also connected to other sites and/or third-party digital information providers to provide personalized and interconnected experiences. For example, individuals reading news stories, shopping online, or listening to music are now connected, which allows instantaneous feedback from trusted sources, immediate conversation, and crowdsourcing of related or tangential information [31]. While many other single sign-on systems are leveraged within enterprise approaches, the socially based versions are mostly based on the Facebook system or through Google's similar connectivity.

The digital identification necessary to bridge the so-called digital divide is not limited to the technological pieces of identification. It also relates to how individuals perceive themselves in a world filled with social media systems, mobile and portable technological devices, and a growing trend toward increasingly more information sharing. However, if the availability of information and devices is not convincing, the presence of these systems and devices during situations where information is not critical clearly indicates that people identify their existence with the ability to access and process information this way. For example, a 2013 Jumio survey found that 72% of Americans say they are within 5 feet of their smartphones the majority of the time, which allows them to regularly use their devices in movie theaters (35% of the time), during a dinner date (33% of the time), at a child's school function (32% of the time), in the shower (12% of the time), and even during sexual activities (9% of the time) [32]. There are no other technologies over the history of time that have had the level of integration and prevalence not only throughout society, but also in the day-to-day lives of their users.

The final component of the modern digital divide is related to ability of an individual to create and manipulate data within these social media and digital systems. Pew Internet refers to those who actively and routinely engage in social media systems as creators, curators, and power users. Specifically, 46% of adult Internet users are creators who post original photos or videos on social media systems. This contrasts to the 41% of adult Internet users who repost photos that were found online or shared with them on social media systems. Overall, 56% of all users were either a creator or curator and 32% of users engaged in both creating and curating activities [33]. Likewise, most Facebook users get more from their friends on Facebook than they give. This phenomenon occurs because of so-called power users who contribute much more than an average user. According to a recent study, between 20% and 30% of Facebook users (depending on activity) were designated as power users due to their much higher daily or weekly engagement. Moreover, these power users are often specialized in a particular activity like sending friend requests, pressing "like" buttons, and tagging friends in photos [34].

This manipulation of data is certainly not limited to Facebook. Similar scenarios occur on Twitter, YouTube, Instagram, and many other social media systems. Likewise, there are many early adopters of digital devices who are willing to wait in long lines and pay premium prices to purchase cutting-edge technologies (e.g., new iPhones). These early adopters of technology are often described as forward-leaning and adventurous consumers; however, they do play a vital role in the growth, implementation, and later integration of technology into broader society. Specifically, they test systems as individuals and crowd collectives to identify strengths and weaknesses and begin to create interest and energy in later adopters. To technology adopters, this process is established through a higher competency of system interaction, acceptance and use of new systematic language, and the creation of so-called cultural capital where people embrace the early knowledge and understanding that comes from early adoptions [35].

Even though traditional research has indicated that only 13.5% of technological consumers are actual early adopters, a recent Harris Interactive poll found that nearly 56% of surveyed adults identified themselves as an early user and adopter of technologies [35,36]. This phenomenon is related to the fact that "the Internet has democratized the culture of early adoption…[so that] being first is no longer reserved for diehard fan boys" [36]. What previously was a relatively steady distribution between early and late adopters has now been flattened with the increase of early adopters of emerging technologies. This shift toward early or first implementers is contributing to addressing this component of the digital divide.

IMPACTS TO EMERGENCY MANAGEMENT AND RESPONSE

Widespread access to the Internet and nearly ubiquitous exchange of information via social media systems and mobile devices have clearly changed when, where, and how people access information. However, this change in the access and exchange of information does not necessarily change any cultural, ethical, or societal norms within given communities. Understanding this dynamic is important to understand how social media and other emerging digital technologies have impacted previous disasters and how they may progress in the future.

In Other Words…Changing Human Nature

While the internet may be changing the way we organize our thinking, and while it is changing the way we organize our relationship with one another, it certainly does not change basic human nature… [but] good and evil…will play out in new ways.

~Former Vice President Al Gore, *The Future: Six Drivers of Global Change* [37]

The influence of social media on disasters and emergency management has a short history. With the establishment of most active social media systems in the mid-2000s (e.g., Facebook in 2004) and the meteoric rise in popularity and growth, the impact of social media and digital systems can be broken down into three phases: public usage, nongovernmental organization (NGO) usage, and governmental usage. While there is some overlap in these phases, these roughly represent the states of consideration, use, adoption, and eventual acceptance of social media systems within disaster preparedness, response, recovery, and mitigation.

The first phase of how social media systems impacted disaster was simply the public usage. Starting as early as the London bombings (2005) and continuing up to and including the Mumbai terrorist attacks (2008), early adopters of social media systems like blogs, Facebook, and Twitter began to post timely and poignant messages and content (e.g., photos) to share disaster-related information. For example, in 2005 during the bombings of the London underground system, active bloggers posted pictures and first-hand accounts that quickly undermined official government reports that the explosions were not acts of terrorism, but instead were caused by a utility system failure. Likewise, in 2007, students utilized social networking sites to confirm the names of all 32 victims of the Virginia Tech campus shooting before any confirmation from the university was provided to the general public. The last major event in this phase was the 2008 terrorist attack in the financial district of Mumbai, India. This event represented the first time documentation about an emergent event was first reported on Twitter rather than on traditional or cable news networks [38]. In all three cases and in similar events, the use of social media systems was solely by citizens. Formal volunteer groups and government entities more often than not ignored these complications or merely reacted retroactively to the implications presenting themselves.

While this level of public involvement in social media systems has only continued to increase, the next phase of utilization was the application of social media and related digital systems by organized and spontaneous groups of volunteers. For example, events such as the Haiti earthquake (2010), the EF-5 Joplin, Missouri, tornado (2011) (see Figure 1.7), and the response to the widespread Alabama tornadoes (2011) were groundbreaking in presence of social media within volunteer systems. For example, groups like CrisisCommons, Mission 4636, and Ushahidi (among others) brought together volunteers from around the globe through digital systems. This rise of crowdsourcing happened again in Joplin and Alabama and has become a common occurrence in most significant emergent events [38]. The significance of this change is that emergency management has traditionally depended on geographically dependent volunteers

Figure 1.7 In response to widespread damage like this from the 2011 Joplin, Missouri, tornado, there was widespread use of social media to provide supplementary support to formalized systems. (Source: FEMA/Jace Anderson.)

that have a natural and organic connection to the impacted community. However, with the rise of social media, this dependency on local resources and all related challenges are now altered.

Unfortunately, government's use and acceptance of social media usage before, during, and after disaster were the last to develop. However, this began to change as early as the earthquake and tsunami in Japan that lead to the Fukushima nuclear facility meltdown in 2011. During this event, social media was utilized by citizens and volunteer groups, representing one of the first major events that governmental operations specifically promoted and utilized social media as a primary communication system. Specifically, within 24 hours of the earthquake striking off the coast of Japan, the US Embassy in Tokyo released a message encouraging Americans in the impacted area to use social media, Google Person Finder, and cell phones to access and share critical information with friends and family in the impacted area and throughout the world [39]. This type of application was repeated during Superstorm Sandy in 2012 when nearly all impacted local, state, and federal emergency management agencies utilized social media to send, receive, and process information during the event.

This short history of social media and disasters is moving at a quick, but predictable pace especially considering that individuals throughout the world have utilized systems for personal use and have leveraged them during a disaster. This includes the systems that impact them personally or the broader community they are engaged with. However, the simple exchange of formal and informal information on social media systems is not the limit of their impact to professional emergency management and response professionals. These additional issue impacts are broad in scope, but mostly relate to operational support, intelligence gathering, and resource availability and management.

OPERATIONAL SUPPORT AND INTELLIGENCE GATHERING

The first emerging characteristic is the impact of social media and digital technologies on operational support systems. In most historical examples, social media monitoring and support have been assigned to public information groups to monitor and distribute official information. Unfortunately, traditional and national incident management systems like the Incident Command System (ICS) and National Incident Management System (NIMS) have difficulty integrating social media (and all of its implications) into best-practice management systems that were developed well before social media became functional to use before, during, or after a disaster. Consequently, review and approval of information is often a time-consuming or organic process that may or may not be effectively utilized for social media. Likewise, it is rare for operational specialists to effectively aggregate, analyze, and apply information or intelligence provided via social media channels. As established earlier in the chapter, nearly every emergent event will generate social media information critical to all phases of response including injuries, fatalities, damage assessment, infrastructure damage, and many others, and it is certainly not limited to public information.

However, these challenges also present unique opportunities to improve response systems if social media information can be leveraged appropriately. For example, many regional, state, and international emergency management organizations are adopting a model developed in 2011 called the virtual operations support team (VOST). Much like the emergent volunteers that rose in prominence after the earthquake in Haiti, the VOST uses non-geographic-specific volunteers in an organized and controlled fashion to supplement local operational resources. This group is then dedicated to the monitoring, aggregation, and application of social media information. These VOST groups integrate into operations centers or command groups at the will and direction of the impacted area. This is a potentially tremendous benefit to the resource-challenged impacted community and should significantly improve the information and intelligence available during disasters. While VOST is not the only model, the future clearly dictates that social media must be harvested for information just like any other source of information to ensure the efficient and effective application of resources during events.

In addition to the significant potential for operational readiness and intelligence improvements, digital technology systems also improve the resource redundancy to local response agencies. For example, cloud-based systems allow for a significant increase in the reliability and restoration of essential functions within the organization or community. For example, in the past critical computer files would have been maintained locally or at best in a locally maintained data center. While these systems can maintain duplicate records or files, the recovery sites are often localized and device based (e.g., tapes or drives), which leads to increased vulnerability to localized events and hazards as well as delayed recovery processes. In contrast, cloud-based solutions inherently address some of these issues through multiple off-site nonlocalized data recovery sites that can quickly be accessed via online systems and mobile devices and are not impacted by local events.

While cloud-based solutions are underutilized by most professional emergency management organizations, they are widely leveraged by businesses and emergent volunteerism groups as the threshold to utilize is much lower and the benefits are significant. As cloud-based solutions continue to increase in use and prevalence, the financial and operational efficiency will be too much not to apply before, during, and after disasters. For example, it is highly likely that all operational computer systems (e.g., incident management systems, emergency notification systems, etc.) will shift away from software or server-based protocols toward off-site cloud-based management options. Likewise, some organizations, like the VOST groups discussed earlier, actively utilize cloud-based document generators (e.g., Google Docs) to conduct simultaneous and real-time editing by multiple users. This type of functionality is not often utilized by emergency public information or operational functions in emergency operations centers (EOCs), but will quickly become commonplace as comfort levels with these tools increase within response communities.

Although unlikely to occur as quickly as data management, fundamental office tools such as word processing, spreadsheets, and basic data management will also be implemented using cloud-based solutions. This projection is aided by the significant growth of cloud-based solutions in this market, including OpenOffice and Google Suite products (e.g., Google Docs). Interestingly, the usage of Google Docs and Google Spreadsheets stayed relatively static within the first 12 months of its release in 2006. However, starting in late 2007, the rate of users on these systems increased nearly 700% over the next year to nearly 1.4 million monthly users [40]. At that time, Google Docs was only used as an enterprise product in 5% of workplaces; however, by 2009 this figure jumped to nearly 20% [41]. Since that time, the institution of Google products as an official enterprise source has increased exponentially. For example, Google reports that 58% of all Fortune 500 companies and 72 of the top 100 universities in the United States have now formally adopted Google's cloud-based productivity and storage solutions [42]. While this is not intended to be a sales pitch for Google, it is important to establish that third-party cloud solutions are clearly shifting previous organizational practices.

THE STRUGGLE WITH TRANSPARENCY

As social media systems have become common and greater access to technology has become commonplace, there is a growing expectation from the general public that local, state, and federal governments will engage in these systems at the same rate and level of openness that members of the public do. This expectation of openness has translated to an open government and transparency movement that has become popular among more progressive government leaders. This is evident in the fact that nearly every state government maintains websites to promote programmatic transparency within their operations for both governance and financial decisions. Likewise, President Obama distributed a memorandum in 2009 (see Figure 1.8), not long after his first election, that stated his administration was "committed to creating an unprecedented level of openness in Government…to ensure the public trust and establish a system of transparency, public participation and collaboration…[to] strengthen our democracy and promote efficiency and effectiveness in Government" [43].

Ironically, this commitment to transparency in government has been more rhetoric than reality in some instances. There has been significant pushback against the US government since President Barack Obama's

Figure 1.8 President Barack Obama issued a memorandum in 2009 calling for the federal government to adopt a widespread philosophy of transparency with strong citizen involvement. (Source: White House/Pete Souza.)

initial declaration for the need to create transparency in regard to how the government dealt with sensitive and classified information as well as the processing of freedom of information (FOI) requests. For example, the rate of classifying government documents remains higher than the declassification rate even though specific and targeted policies have been suggested. Likewise, the American Civil Liberties Union (ACLU) sued the US federal government to release more information about its wartime use of drones [44]. On a similar but less serious note, President Obama was given an award on March 28, 2011, by a coalition of open-government advocates to honor his administration's commitment to transparency; however, the meeting was closed to reporters and photographers and was not announced on the president's public schedule [44].

In the realm of disaster management there are even fewer strong attempts at transparency. In most cases, the recovery process is the most common phase where openness can easily occur, given the necessity of widespread involvement of local, regional, state, federal, and private assets to restore community needs. Perhaps more interesting is the use of openness by highly structured response agencies like the US National Transportation and Security Board (NTSB). The primary function of the NTSB is to investigate airplane, train, and other major transportation accidents that happen. Because of the highly complex and often sensitive nature of these events, there was significant attention given to the NTSB for their significant use of Twitter and other social media outlets to document their response following the crash of Asiana flight 214 that crashed at the San Francisco airport in 2013. While some airline unions objected to this highly open release of data, claiming that it created confusion and doubt in the general public, the NTSB simply committed to using the same tools used by the public to allow the "wholesale transfer of once-obscured data" [45]. This was ultimately widely supported by Twitter users including one who commented: "In the age of Twitter, I don't think it makes sense to withhold any findings at all until 9 months down the road… A large plane crashed at a major American airport…[and] people want to know why and whether the cause might be relevant to their own future flying" [45].

SUPER DIGITAL AGE

The developments discussed throughout this chapter merely established the brief history of social media and disasters and begin to consider the future. Unlike when some trailblazing professional emergency managers began using social media in 2008, there is no indication that social media or digital technologies have peaked, plateaued, regressed, or become a fad in any way. While prediction (as discussed in the introduction) is never 100% accurate, it is reasonable and fair for emergency managers to continue to embrace the conceptual impacts of when, where, and how these systems can be leveraged to reduce the impacts to communities impacted by emergencies or disasters.

While nearly all professional emergency management disciplines have begun to utilize social media for the distribution of information, there are very few that have truly leveraged these same systems to the full capability as the general public has. At this point, the use of social media during emergencies or disasters is commonly just an updated form of a press release that simply pushes information and never seeks to engage in the audience by seeking, confirming, or otherwise leveraging the awareness or capabilities of the digital community. This lack of integration into operational systems and protocols is consequently the area where there needs to be the most significant growth in the years to come. Model practices like VOSTs or something similar will eventually become commonplace for most communities. Moreover, the professional emergency management field must also make philosophical changes to how it considers emerging tools. Rather than sluggishly adopting these newer technologies when there is no choice but to do so, emergency managers and professional responders must be less concerned about the limitations and more focused on the possibilities for success and integration into the communities they represent.

REFERENCES

1. Thompson, Clive. (2013). "Vision Quest." *Wired*. September 2013.
2. Holmes, Thomas. (2012). "History of Tornado Sirens." Ensuring Safety through a Sound Blog. http://blogs.creighton.edu/trh42834/2012/05/31/history-of-the-tornado-siren/. Accessed March 15, 2014.
3. Rasmaussen, Cecilia. (2007). "Long Silenced Air-Raid Sirens Are Relics from a Jittery Past." *LA Times*. April 29, 2007. http://articles.latimes.com/2007/apr/29/local/me-then29. Accessed March 15, 2014.
4. Crowe, Adam. (2013). *Leadership in the Open: A New Paradigm in Emergency Management*. Boca Raton, FL: CRC Press.
5. Erickson, Christine. (2012). "A Brief History of Text Messaging." *Mashable*. http://mashable.com/2012/09/21/text-messaging-history. Accessed March 16, 2014.
6. Reardon, Margaurite. (2008). "Americans Text More Than They Talk." CNET. http://news.cnet.com/8301-1035_3-10048257-94.html. Accessed March 16, 2014.
7. "Mobile Technology Fact Sheet." (2013). Pew Internet Research Project. http://www.pewinternet.org/fact-sheets/mobile-technology-fact-sheet. Accessed March 16, 2014.
8. McMillian, Robert. (2013). "The Friendster Autopsy: How a Social Network Dies." *Wired*. http://www.wired.com/wiredenterprise/2013/02/friendster-autopsy/. Accessed March 17, 2014.
9. "Key Facts." (2014). Facebook Newsroom. http://newsroom.fb.com/Key-Facts. Accessed March 17, 2014.
10. "Twitter Statistics." (2014). Statistics Brain. http://www.statisticbrain.com/twitter-statistics/. Accessed March 17, 2014.
11. "Number of Active Users at Facebook over the Years." (2013). *Yahoo News*. http://news.yahoo.com/number-active-users-facebook-over-230449748.html. Accessed March 17, 2014.

12. "Over the Years: Twitter's Growth in Active Users." *Business Today.* http://businesstoday.intoday.in/story/over-the-years-twitter-growth-in-monthly-active-users/1/199237.html. Accessed March 17, 2014.
13. Buck, Stephanie. (2013). "Cell-ebration! 40 Years of Cell Phone History." *Mashable.* http://mashable.com/2013/04/03/anniversary-of-cellphone. Accessed March 19, 2014.
14. "Mobile Technology Data Sheet." (2014). Pew Research Internet Project. http://www.pewinternet.org/fact-sheets/mobile-technology-fact-sheet/. Accessed March 19, 2014.
15. "Infographic: 2013 Mobile Growth Statistics." (2013). Digital Buzz Blog. http://www.digitalbuzzblog.com/infographic-2013-mobile-growth-statistics/. Accessed March 19, 2014.
16. "Facebook Turns 10: Key Facts to Remember." (2014). *Social Media Today.* http://socialmediatoday.com/gonzogonzo/2136276/facebook-turns-10-ten-facts-consider. Accessed March 20, 2014.
17. Wholsen, Marcus. (2014). "The PC's Death Might Also Mean the Web's Demise." *Wired.* http://www.wired.com/business/2014/01/death-pc-also-mean-end-web/?cid=16983574. Accessed March 20, 2014.
18. Rapaport, Richard. (2014). "A Short History of the Digital Divide." Edutopia. http://www.edutopia.org/digital-generation-divide-connectivity. Accessed March 20, 2014.
19. Strover, Sharon. (2003). "Remapping the Digital Divide." *The Information Society.* http://www.indiana.edu/~tisj/readers/full-text/19-4%20guest.pdf. Accessed March 21, 2014.
20. Guzman, Monica. (2013). "Four Digital Divides: Where Do You Stand?" *Seattle Times.* http://blogs.seattletimes.com/monica-guzman/2013/11/02/four-digital-divides-where-do-you-stand/?utm_content=bufferaa202&utm_source=buffer&utm_medium=twitter&utm_campaign=Buffer. Accessed March 21, 2014.
21. Fox, Zoe. (2013). "5 Billion People Will Use Mobile Phones by 2017." *Mashable.* http://mashable.com/2013/10/03/mobile-phones-2017. Accessed March 21, 2014.
22. "Lifeline Program for Low Income Consumers." (2014). Federal Communications Commission. http://www.fcc.gov/lifeline. Accessed March 23, 2014.
23. "Lifeline: What Is Support?" (2014). Universal Service Administrative Company (USAC). http://www.usac.org/li/telecom-carriers/step01/default.aspx. Accessed March 24, 2014.
24. Meeks, Brook N. (2000). "Is Privacy Possible in the Digital Age?" MSNBC Opinion. http://www.nbcnews.com/id/3078854/t/privacy-possible-digital-age/#.UzDl0_ldWSo. Accessed March 24, 2014.
25. Bullas, Jeff. (2012). "35 Mind-Numbing YouTube Facts, Figures, and Statistics." Jeff Bullas Blog. http://www.jeffbullas.com/2012/05/23/35-mind-numbing-youtube-facts-figures-and-statistics-infographic/. Accessed March 24, 2014.

26. "Facebook Statistics." (2014). Statistics Brain. http://www.statisticbrain.com/facebook-statistics/. Accessed March 25, 2014.
27. Madden, Mary et al. (2013). "Teens, Social Media, and Privacy." Pew Internet. http://www.pewinternet.org/2013/05/21/teens-social-media-and-privacy/. Accessed March 25, 2014.
28. "Edward Snowden Biography." (2013). Biography. http://www.biography.com/people/edward-snowden-21262897?page=2. Accessed on March 25, 2014.
29. Temple-Raston, Dina. (2013). "Secrets Just as Hard to Maintain as Privacy in Digital Age." NPR's All Things Considered. http://www.npr.org/templates/story/story.php?storyId=190756384. Accessed on March 25, 2014.
30. Abelson, Hal and Lessig, Lawrence. (1998). "Digital Identity in Cyberspace." White paper. http://groups.csail.mit.edu/mac/classes/6.805/student-papers/fall98-papers/identity/linked-white-paper.html. Accessed on March 25, 2014.
31. Axon, Samuel. (2010). "Facebook's Open Graph Personalizes the Web." *Mashable*. http://mashable.com/2010/04/21/facebook-open-graph. Accessed on March 27, 2014.
32. "Where Do You Take Your Phone?" (2013). Jumio Blog. http://www.jumio.com/2013/07/where-do-you-take-your-phone. Accessed on March 27, 2014.
33. "Social Networking Fact Sheet." (2013). Pew Research Internet Project. http://www.pewinternet.org/fact-sheets/social-networking-fact-sheet/. Accessed on March 27, 2014.
34. Hampton, Keith et al. (2012). "Why Most Facebook Users Get More Than They Give." Pew Research Internet Project. http://www.pewinternet.org/2012/02/03/why-most-facebook-users-get-more-than-they-give. Accessed March 27, 2014.
35. Gerard, Morgan. (2014). "Innovations and Early Adopters: Beyond the Bell Curve." NoodlePlay Blog. http://www.ideacouture.com/blog/innovation-early-adopters-beyond-the-bell-curve. Accessed March 28, 2014.
36. Soper, Taylor. (2014). "Study: Majority of U.S. Adults Call Themselves Early Adopters of New Technologies." GeekWire. http://www.geekwire.com/2014/study-majority-americans-now-early-adopters-new-technology/. Accessed March 28, 2014.
37. Gore, Al. (2013). *The Future: Six Drivers of Global Change*. New York: Random House. p. 69.
38. Crowe, Adam. (2012). *Disasters 2.0: The Application of Social Media in Modern Emergency Management*. Boca Raton, FL: CRC Press.
39. "How to Contact the U.S. Embassy, Your Friends and Family." (2011). U.S. Embassy—Tokyo, Japan. http://japan.usembassy.gov/e/acs/tacs-warden20110312-03.html. Accessed March 29, 2014.
40. McManus, Richard. (2007). "Google Docs and Spreadsheet Usage Increasing—But Is Google Apps the Market Leader?" ReadWrite Blog. http://readwrite.com/2007/12/06/google_docs_spreadsheets_usage_increasing#awesm=~oA8SFEI04MwnVG. Accessed March 31, 2014.

41. Thibodeau, Patrick. (2009). "Google Docs 'Widely Used' At 1 in 5 Workplaces, Survey Finds." *Computer World*. http://www.computerworld.com/s/article/9138157/Google_Docs_widely_used_at_1_in_5_workplaces_survey_reports. Accessed March 31, 2014.
42. "Google Enterprise." (2014). Google. http://www.google.com/enterprise/. Accessed March 31, 2014.
43. "Transparency and Open Government." (2009). The White House. http://www.whitehouse.gov/the_press_office/TransparencyandOpenGovernment. Accessed April 1, 2014.
44. "The Best Disinfectant." (2012). *The Economist*. http://www.economist.com/node/21555924. Accessed April 1, 2014.
45. Bachman, Justin. (2013). "In Probing the Asiana Crash, NTSB Gets Busy on Twitter." *BusinessWeek*. http://www.businessweek.com/articles/2013-07-11/in-probing-the-asiana-crash-ntsb-gets-busy-on-twitter. Accessed April 1, 2014.

2

Communication and Engagement

Giants are not what we think they are. The same qualities that appear to give them strength are often the sources of great weakness.

~Malcolm Gladwell, *David and Goliath* [1]

TELEVISION, RADIO, AND PRINT MEDIA

Communications is a critical element of professional emergency management operations. Information exchanged before, during, and after disasters—whether formally, informally, technically, or organically—ensures decisions and choices are made that reduce the threat to those individuals, facilities, and community components that may have been impacted by an emergent event. However, one of the most common challenges identified by professional emergency managers, after exercises, training sessions, and real events, is a failure of communications. These failures can occur for a variety of reasons including infrastructure failures, political sensitivities, diversity of systems, personality conflicts, and unmet expectations. This chapter will look at how communication processes and corresponding engagement have changed over the last decade and what trends exist that emergency managers can use to forecast and predict communication needs into the future.

In the broadest sense, there are five major sources of information for the general public: print, radio, television, Internet, and, most recently, social media. Print, radio, and television media are typically considered to be or referred to as "traditional media" as most of these forms of communication and information distribution are often primarily associated with professional organized news reporting agencies or organizations. Likewise, while there is significant overlap between these communication media categories, they provide information in vastly different ways and often serve completely different audiences.

Print media forms are the oldest and most well established of the five communication types. All print media originated with Johannes Gutenberg's invention of the handpress in the 1450s. According to one scholar, Gutenberg's press was "in some respects less an 'invention' than it was a clever synthesis of existing technologies" [2]. Within decades, Gutenberg's combination of a rapid hand-powered screw press and movable type had spread through Germany and much of Western Europe (see Figure 2.1). This handpress design remained nearly unchanged for

Figure 2.1 Johannes Gutenberg's press served as the initiator of one of the oldest and most well established forms of traditional media. (Source: From Die großen Deutschen im Bilde.)

more than 350 years, but by the 1800s the rise of industrial technologies allowed for larger printed surfaces and an increase in the volume of printed material [2]. While the printing process has certainly been modernized in production, the fundamental structure and layout have not. Newspapers and magazines have existed in the same basic format for a century.

Because of the maturity of printed media, newspapers have been a primary source of disaster-related information for the vast majority of disasters that occurred before the rise of television news and later cable television news. Major disasters or global events such as the sinking of the *Titanic* (1912), dropping of the atomic bomb on Hiroshima and Nagasaki (1945), starting of wars, and assassinations of world leaders were all announced to the world via newspapers (see Figure 2.2). The traditional

"All the News That's Fit to Print."

The New York Times.

TITANIC SINKS FOUR HOURS AFTER HITTING ICEBERG;
866 RESCUED BY CARPATHIA, PROBABLY 1250 PERISH;
ISMAY SAFE, MRS. ASTOR MAYBE, NOTED NAMES MISSING

Col. Astor and Bride, Isidor Straus and Wife, and Maj. Butt Aboard.

"RULE OF SEA" FOLLOWED

Biggest Liner Plunges to the Bottom at 2:20 A.M.

RESCUERS THERE TOO LATE

The Lost Titanic Being Towed Out of Belfast Harbor.

PARTIAL LIST OF THE SAVED.

CAPT. E. J. SMITH.

Figure 2.2 Major disaster events like the sinking of the *RMS Titanic* were first reported in newspapers or other print media. (Source: *The New York Times.*)

power and influence of newspapers was particularly evident in 1898 when the *New York Journal*, in an act of so-called "yellow journalism," blamed Spain for a mysterious explosion that sank the *USS Maine* in Havana Harbor, Cuba, and ultimately helped push the United States into war with Spain [3].

Unfortunately, the influence of newspapers has been on the decline since the 1950s and most likely will fade away completely as the faster and more instantaneous forms of information (e.g., social media and Internet) rise in prominence and use. Specifically, in the 1950s the total number of paid newspaper subscriptions in the United States equaled roughly the total number of households. While the number of subscriptions rose slightly until 1990, it fell far behind the steady growth of households over that same period as additional media forms—particularly television—grew in popularity. After 1990 the number of newspaper subscriptions fell steadily and by 2010 the number of subscriptions fell to roughly the same level as in 1955 [4]. Moreover, while some would argue that this reduction in newspaper readership was predominantly from the changes or closure of smaller, less profitable newspapers, the subscription rates for major American newspapers like the *Los Angeles Times*, the *Washington Post*, and the *Daily News* all fell by nearly 50% between 1990 and 2010 [5].

This precipitous drop was caused by a variety of reasons that nearly all related to the invention of the Internet and the rise of social media. Specifically, newspaper revenue has traditionally been driven by paper sales, advertisements, and personal "want" ads. However, paper sales declined as the relevancy of newspaper information became dated (published no more than daily) and personal want ads were available for free on Internet sites like Craigslist (see later discussion). Consequently, with fewer readers, the advertising rates and demand declined in turn. Clearly, the Internet is providing the same information in a quicker, cheaper, and more easily accessible format. This issue is further exacerbated by the use of mobile devices to send, share, and access information in a nearly instantaneous fashion when compared to newspapers that are printed at most once per day.

In Other Words...Why Newspapers Are Dying

A newspaper is a package of content—politics, sports, share prices, weather and so forth—which exists to attract eyeballs to advertisements. Unfortunately for newspapers, the internet is better at delivering some of that than paper is. It is easier to search through job and property listings on the web, so classified advertising and its associated revenue is migrating onto the internet. Some content, too, works better on the internet—news and share prices can be more frequently updated, weather can be more geographically specific—so readers are migrating too.

~*The Economist* (2009) [4]

This impact is most clearly evident through the impact of Craigslist on the use and impact of classified advertisements in newspapers. As one of the major revenue-generating sources for newspapers (along with magazines), classified ads have long been used by local community members to post smaller items (e.g., furniture or used automobiles) sold from one person to another via a private transaction that is facilitated via the newspaper ad. To facilitate these connections, individuals often paid incremental fees (e.g., $25) based on the length or space needed for the ad. Interestingly, a 1964 book by Marshall McLuhan called *Understanding Media: The Extensions of Man* stated that "the classified ads…are the bedrock of the press…[and] should an alternative source of easy access to such diverse daily information be found, the press will fold" [4]. This alternative source—predicted by McLuhan—arrived in 1996 in the form of a basic website created by a computer programmer (named Craig) simply trying to help his friends and their friends by allowing information—from both seekers and sellers—to be posted for free [6]. Since that time, Craigslist has grown to 700 localized sites, 50 million daily queries, and 40 million monthly postings [7]. That level of self-controlled, independently created, and highly affordable format has vastly contributed to the current death march of newspapers.

This trend has significantly impacted how emergency public information officers and disaster managers consider the distribution of information. Basic and advanced disaster public information courses still teach the necessity of well-written and -crafted press releases that are disseminated to print media to announce critical information during emergencies or disasters. Under best-practices models like the Incident Command System (ICS) and the National Incident Management System (NIMS), these press releases must be reviewed and approved by the incident commander or emergency operations center (EOC) manager to ensure there is clarity and consistency of information. Unfortunately, the use of social media by citizens and the exchange of information between direct observers of the event and those who need information have repeatedly been shown to be much faster and often more poignant than officially released information.

Newspapers are not the only traditional media form that has been significantly impacted. For example, radio broadcasting grew significantly in the 1920s. Even though the technology was at its infancy at that point, audiences quickly determined that radio was a faster means of receiving updated information than printed newspapers were. Events such as the 1920 Harding–Cox presidential election and the 1925 Scopes "Monkey Trial" were early examples of events that were first announced via radio systems [8] (see Figure 2.3). With increasing relevancy, the number of radio sets and ultimately radio broadcasting stations increased significantly. For example, there were 28 stations in operation in the United States in 1922, but this number jumped to 1,400 by 1924. Some of the earliest stations were the National Broadcasting Company and the Columbia Broadcasting System, which are still familiar as television networks NBC and CBS, respectively [8].

While the percentage of people listening to AM and FM radio each week has remained around 90% for the last decade, the use of radio broadcasting to receive emergency news information has declined. Specifically, there was a 40% reduction in the number of individuals utilizing radio to receive this type of information. While this shift did coincide with an increase in those seeking information via the Internet, the decline in radio usage started well before emergency news information was

Figure 2.3 Radio was traditionally used for families to receive information about major events such as emergencies and disasters. (Source: Library of Congress.)

readily available via online outlets or social media systems [9]. While it is not wholly clear why this decline has occurred, there are several prime considerations that are particularly plausible as technology continues to emerge and culture and communication change around it.

The first major challenge to radio as an emergency information source is the diversity of options. For example, there is a growing divide between so-called "terrestrial" radio (e.g., traditional AM/FM stations played via radios), satellite radio, and Internet radio. For example, online streaming radio listeners have steadily increased since 2002 with weekly listeners at 29% and monthly listeners at 39% when last measured in 2012 [9]. While this is still significantly lower than for traditional radio, it shows a significant and steady additional and/or optional choice for information. Online streaming radio systems are often provided in formats with few interruptions from mandated station identifications or advertisements that are commonplace in terrestrial radio systems. Moreover, online streaming radio systems are also widely available through mobile phone apps and thus can be accessed without limit to the location of the radio or proximity to the radio signal tower. Specifically, nearly 20% of online radio listeners have done so in their automobiles via mobile phones, which is a nearly 100% increase annually since 2010 [9].

Another significant challenge to the traditional radio format is the sheer number and diversity of stations and audio formats. For example, according to a recent State of the News Media Report released by the Pew Research Center, there are more than 4,000 traditional radio stations in the United States alone. Of those stations, there are 11 different subcategories of format with no classification constituting more than 15% of the stations. Specifically, the fewest number of stations were those designated as "Mexican regional," which constituted only 3% of all radio stations. Conversely, those stations designated "country/new country" represented the largest number of stations at 14.1%. Just below "country" stations were those listed as "news/talk/information/personality," which constituted merely 12.1% of the stations. Alternatively, it could be said that nearly 88% of the radio stations do not focus on emergent issues and do not provide formats that would easily support information during emergency or disasters. Likewise, a growing number of stations (particularly those not categorized as news related) are not controlled locally. For example, during certain nonpeak hours or holidays, the programming of corporate-owned stations is run from central engineering stations that are not necessarily in the local market. As such, it is nearly (if not totally) impossible to share emergent and critical information via these radio stations during emergencies and disasters.

In addition to the challenge of traditional choices in terrestrial radio, online streaming radio has also created a significant impact to when, where, and how people receive information on the go. With the universal addition to AM/FM radios to vehicles in the 1950s, people most often have accessed radio information while on the go with preselected stations that serviced a local community based on a common geography around the station [10]. This geographic range was set (and often limited) by the power of the station transmitter. However, the availability of portable music devices, satellite radio, and mobile phones has shifted the use and purposes of transported information.

For example, according to a recent State of the Media survey, online car listening through cell phones jumped from 6% to 17% between 2010 and 2012 [9]. Likewise, satellite radio rose from 20% to 24% over that same period of time [9]. This trend toward online streaming radio will only get more prominent in the future as streaming services like Pandora already reach more than 40% of the 18- to 24-year-old demographic [11]. Much like traditional terrestrial radio stations, streaming audio services are often on-demand in the mix or station availability, which exponentially increases the total number of options available and eliminates the geographic limits and therefore focus of terrestrial radio. Consequently, the capability to push relevant and local information is becoming increasingly less effective with no forecast of change in the near future (see Figure 2.4).

Figure 2.4 Traditional radio has faced challenges to the effectiveness and reach as individuals shift listening and information choices. (Source: FEMA/George Armstrong.)

Because of these trends of shifting away from radio and toward non-geographic-dependent listening options, some of the traditional emergency notification strategies have also begun to be modified. For example, the Emergency Alert System (EAS) replaced the Emergency Broadcasting System (EBS) in 1997 as a means for a national public warning system that would utilize (via required participation) local television and radio broadcasters as well as cable and wireless cable systems to offer emergency communications to locally impacted or national emergencies [12,13]. The activation of this system falls under the jurisdiction of approved local authorities, the National Weather Service, or the president (under certain circumstances). This system also provides for the backbone of the NOAA All-Hazard Radio system that can actively alert people based on preselected geographic characteristics. Unfortunately, these emergency systems have reduced impact if people are not utilizing systems like streaming radio that are not connected to the service and lack the geographic framework to provide clear and intentional warnings to the impacted areas.

To address these gaps and maximize the growing number of mobile device users, the US Department of Homeland Security (DHS) in conjunction with the National Weather Service and major cellular phone service providers (e.g., AT&T and Verizon) has implemented wireless emergency alerts (WEAs) through the Integrated Public Alerting and Warning System (IPAWS). Since its full rollout in 2012, wireless emergency alerts have been sent out throughout the United States to alert local communities about flooding, evacuation routes, and impending tornadoes as well as Amber Alerts, which helped safely recover abducted children [14]. The WEA messages are not text messages, but rather pushed mobile messages that utilize a unique technology developed to ensure that message delivery is immediate and avoids potential congestion on typical mobile or wireless networks (see Figure 2.5). Moreover, the WEA system uses a point-to-multipoint system which allows for messages to be sent within a targeted area. For example, if a New York resident were visiting California during an earthquake, his or her phone would still receive a WEA message sent by California-based emergency response organizations [12].

Given this widespread shift toward the WEAs and the growing utilization of nonlocalized streaming and satellite, it is highly likely that traditional radio stations will continue to decrease in effectiveness as an emergency information dissemination system. While this transition is certainly not complete, it does firmly support the concept that traditional

Figure 2.5 Wireless emergency alerts (WEAs) are one of the newest ways to utilize modern technology for emergency notifications. (Source: FEMA/Hans Yu.)

media forms (like the print media already reviewed) have lost effectiveness in most communities to serve as primary sources of information during these emergent events.

Television broadcast news reporting has long been the most significant source of disaster-related information. Television grew and expanded significantly between the 1960s and the 1980s with the transition of programming from black and white to color, and by 1972 roughly 50% of American households had a color television [15]. By 2013, 99% of the households in the United States owned at least one television and the average was 2.24 televisions per family with 65% having three or more [16]. The presence of televisions in American homes is only part of understanding the evaluation of the impact of this traditional media form (see Figure 2.6). For example, the average American spends between 3.25 and 7.25 hours (depending on race and demographics) each day watching television programming. To put this into perspective, an average American child watches 1,200 hours of television programming annually, but only spends 900 hours in a school environment. Moreover, that same child will see more than 150,000 acts of violence on television by the time he or she reaches the age of 18 [16].

Television news can be divided into three major categories: local news, network news, and cable news. Local news is geographically

Figure 2.6 The presence of televisions in American homes has long been viewed as the most highly impactful form of traditional media. (Source: National Archives and Records Administration.)

based—much like terrestrial radio stations—and provides regularly scheduled newscasts as well as breaking news for emergent events. With the exception of 2013, viewership of local news has decreased every year since 2008 [17]. Likewise, cable news channels like Fox News, CNN, and MSNBC lost viewership by a collected 11% in 2013, which was the lowest number of viewers since 2007 [17]. Network (e.g., ABC, CBS, and NBC) news stations have not been as significantly impacted due to continued strong ratings from morning news programming (e.g., NBC's *Today Show*), which incorporate entertainment and cultural information in addition to more traditional news stories.

The period of time evaluated for these various television news forms is significant as it loosely represents the period of time when social media shifted from a new and limited communication form to an international tool for the dissemination of information and establishment of community. For example, Facebook users went from 58 million in 2007 to 1.23 billion by the end of 2013 [18]. As was established in Chapter 1, other social media systems grew over this same period of time in total users, active users, and posted content. While this correlation cannot be fully identified as causation for this decline in viewership for television, it is a connection that should be strongly considered as the behavior and actions of the television news agencies support such a deduction.

While most local television news stations have finally shifted out of the downsizing that occurred through much of the last half-century, there is a significant and noteworthy spike in the number of writing and editorial positions being hired in online and digital news providers. For example, Buzzfeed and Gawker added 170 and 132 editorial positions, respectively, in 2013. Likewise, Mashable and Yahoo News hired high-profile editors from traditional news outlets like the *New York Times* [17]. Moreover, every form of traditional media—but particularly television—utilizes social media to seek out incident information, event validation, and to ultimately push people toward viewership. This social media engagement includes official social media channels on behalf of the respective television channel as well as individual accounts for the on-air journalists and personalities.

RISE OF SOCIAL MEDIA

The use of social media by traditional media outlets did not occur immediately. Many professional journalists and government leaders felt that social media was a communication fad that would pass into history with minimal impact to traditional media outlets. However, starting with the 2008 terrorist attacks in the financial district of Mumbai, India, social media became a primary outlet of emergent information. Specifically, the first information about this event was initially posted on Twitter along with numerous other reports about the actions of the terrorists and the victims [19,20]. Likewise, photo-sharing site Flickr was used by a local man named Vinukumar Ranganathan, who took 112 photos with his camera and shared them online [20]. In both cases, the information and media were available far in advance of traditional media outlets that lacked the initial awareness and later the resources on the ground in the impacted area to serve as an effective disseminator of critical information. This type of emergent event quickly altered the approach of local, network, and cable news outlets interested in continuing to be the primary and most effective source of breaking information and news.

Given that traditional media can no longer be the sole providers of critical information during emergencies and disasters, many traditional media outlets have begun to adopt a model called citizen journalism that allows those individuals closest to the event (e.g., witnesses or survivors) to document and share information as a supplement to traditional reporting. *Mashable* defines citizen journalism as the opportunity for public citizens to plan "an active role in the process of collecting, reporting, analyzing, and disseminating news and information" [21]. The first consideration of this by traditional media was actually after the London bombings in 2005. Several years ahead of widespread use, the BBC received more than 1,000 photographs, 20 pieces of video, 4,000 text messages, and 20,000 e-mails from citizens impacted by the bombings. While their use was limited during this particular event, the BBC summarized the future impact by saying, "…when major events occur, the public can offer us as much new information as we are able to broadcast to them…[and] from now on, news coverage is a partnership" [22]. Since that time, this process has been validated repeatedly during major events such that nearly every major event is first defined by similar contributions from social media.

Perhaps one of the more interesting applications of new relationships between the traditional media and citizen journalism was the application as a collective intelligence system, which is sometimes referred to as crowd investigations. For example, immediately following the Boston Marathon bombings in 2013, federal response agencies like the US Federal Bureau of Investigation (FBI) released information from the internal investigation through traditional media sources (see Figure 2.7). This information included pictures of bomb components including a black backpack, pressure cooker remnants, and a battery and was quickly released by the traditional media via primary mechanisms (e.g., breaking new release) and via official social media channels. In turn, a geographically disconnected group of people from throughout the world came together via online social media systems and communities in a shared and common interest to help professional investigators find the perpetrators of the horrific crime.

For example, literally thousands of posts were shared on an information-sharing site called Reddit in what became the most well-known crowd intelligence operations ever organically conducted. While many interesting connections and additional photos and videos were shared

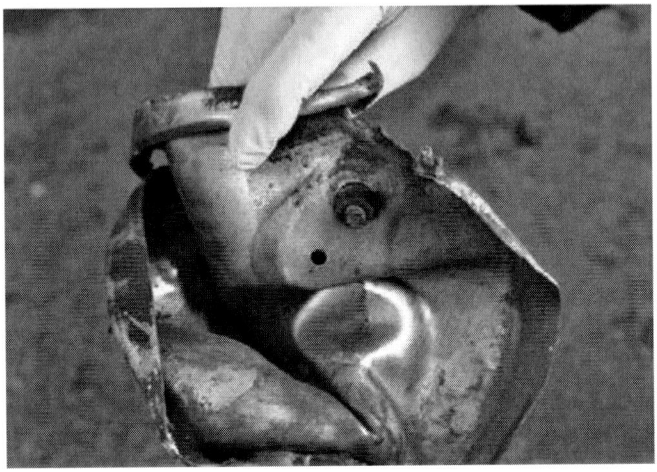

Figure 2.7 Information like the remnants of the pressure cooker bomb was disseminated by the FBI via traditional and social media outlets. (Source: US Federal Bureau of Investigation.)

there, the common deductions were not always correct. The crowd operation on Reddit misidentified several different individuals including the so-called "blue robe guy" and a 22-year-old Brown University student named Sunil Tripathi, who were both highly profiled in traditional and social media [23]. The misidentification of these individuals (and others) was considered so egregious that Reddit's general manager posted an apology that stated, "Activity on Reddit fueled online witch hunts and dangerous speculation which spiraled into very negative consequences for innocent parties…[which] the Reddit staff and the millions of people on Reddit around the world deeply regret" [24].

Interestingly, even though crowd investigations had significant, although unintended, negative implications during the response to the Boston Marathon bombing, it is clear that these sorts of actions will continue to increase as social media and mobile devices continue their trend of becoming increasingly present and relevant to emergency and nonemergency situations. While research is unclear as to why this process is occurring, anecdotal evidence from multiple disasters shows the desire for people impacted by an event to help resolve it in a quick and efficient process. The desire for this process falls somewhere on a scale between narcissistic and altruistic behaviors with some being a mixture.

The altruistic behavior is well defined by social scientists that consistently note that people respond in altruistic and sympathetic ways during emergencies and disasters. These actions are most often seen in the form of spontaneous and unaffiliated volunteers as well as uncontrolled disaster donations and fundraising. This process has not slowed down as text-to-donate campaigns are widely utilized during disaster response to raise millions of dollars. Likewise, emergent volunteerism groups arise after nearly every major disaster to provide support in a nonaffiliated system to those impacted or in need of the event [19]. For example, immediately following the July 2011 bombing and shootings on Norway's Utoya Island, individuals from throughout the area provided private boat launches to rescue victims, unlocked Wi-Fi networks, circulated emergency contact information, and opened homes to survivors and their families [25]. These types of altruistic responses have been documented numerous times for events such as the terrorist attacks of 9/11 and earthquakes, tsunamis, and nuclear meltdowns where "the aftermath of natural disasters…[was] characterized by heroism and a sharing of resources—within the affected community and in others farther away" [26].

In contrast to the constant presence of altruism, many would argue that there has also been a steady increase in the level and pervasiveness of narcissism. This rise directly correlates to the creation of digital social media systems like MySpace, Facebook, Twitter, YouTube, Instagram, and more, which are all built on visual profiles that collect and curate a digital image of a person. Inherently, these digital profiles focus primarily on positive features and values with systematic avoidance of negative features or characteristics. While not a fundamental intention of social media systems, this certainly creates a slippery slope of behavior and communication that can lead toward partial, if not full, narcissism.

In Other Words…Social Media and Narcissism

Needless to say, most social media users are not narcissistic. Yet, social media is to narcissists what crack is to crack addicts: the more narcissistic you are, the heavier your social media use is. Indeed, scientific studies have shown that the number of status updates, attractive selfies, check-ins, followers and friends, are all positively correlated with narcissism, as is the tendency to accept invites from strangers,

> particularly when they are attractive. The reason for these correlations is that narcissistic individuals are much more likely to use social media to portray a desirable, albeit unrealistic self-image.
>
> ~Tomas Chamorro-Premuzic [27]

The use of selfies is not limited to everyday activities. There are an increasing number of so-called "disaster selfies" that are photos taken by a victim or survivor that are quickly shared on personal social media sites. In 2014, there were at least three documented cases of disaster selfies immediately following transportation accidents and major acts of violence. The first noted disaster selfie that gained some traction within social media systems like Twitter and Facebook was taken in January 2014 by a man named Ferdinand Puentes, who had just crashed the Cessna airplane he was flying into the Pacific Ocean off of Molokai, Hawaii. In the picture, Puentes is afloat in the ocean with the plane's tail at a significant angle in the water behind him [28]. Likewise, by March 2014, a Twitter user named Hannah Udren took a selfie photo in front of a US Airways flight out of Philadelphia that nearly crashed on takeoff after it had a nose gear malfunction [28,29]. As a testimony to the power and impact of the disaster selfie, Udren's photo was simply tagged with the words "So Yup" [29].

The last (so far) of infamous disaster selfies was taken by a Franklin (PA) Regional Senior High School student named Nate Scimio, who was credited with helping save lives during a massive stabbing event at his high school. Scimio posted a picture from a local children's hospital that showed a bandage on his arm covering a stab wound he had received during the event. Much like Udren's posting, Scimio's Instagram post included the simple message "Chillin' at Children's [Hospital]" [28]. However, unlike the previous two examples, Scimio's disaster selfie generated significant public commentary about the appropriateness of such postings. For example, contrary to one tweet that attempted to defend Scimio's use of the selfie as a modern communication form, another user commented that Scipio was "taking to social media to get famous from his paper cut wound—pathetic" [30]. Clearly, the acceptance and prevalence of selfies in conjunction with the national focus related to the school shootings led to a unique and special crossroads where new communication forms are emerging that not only inform concerned friends, family, and followers, but also potentially serve as a psychological tool to those directly and indirectly impacted. In short, one commentary stated that "not only is there nothing wrong, nothing inappropriate or nothing tone deaf on the timing of the [disaster] selfie, it's actually a good thing, completely appropriate and a psychologically natural thing to do in the situation" [30].

IMPACTS OF COMMUNICATION SHIFTS

The shift away from solely traditional media sources and the use of digital forms of communication (including social media) have created several significant and distinct changes in the landscape of communications and engagement before, during, and after significantly disruptive events. These impacts include a shift in the perception of time and duration, social accountability, information permanency, information credibility, and physiological changes. Understanding this shifting landscape will help better prepare formal communication systems to adapt to these systems as they change.

The first major shift to be considered is the impact on the perception of time and duration. This shift is based on changes in the amount of information that humans are capable of receiving and processing in a given period of time. For example, a 1980 study estimated that Americans received slightly more than 7 hours of information on an average day (not including work environments). However, by 2009, that amount of information increased to nearly 12 hours of information each day [31]. Likewise, another study estimated that in 2008 Americans consumed nearly 11 trillion words of information over the course of the year [31]. These figures represent an astronomical amount of information that was processed before social media systems became fully ubiquitous and before mobile, digital devices like iPhones were present in millions of hands.

To compensate for the information available via the Internet, social media, and digital applications, the University of San Diego released a report stating that by 2015 the "sum of media asked for and delivered to consumers" would exceed 15 hours a day. That rate of consumption is nearly the equivalent of nine DVDs worth of data per person per day [32]. Of that amount, Facebook and YouTube will constitute 35.2 billion hours annually by 2015 [32]. This amount of information clearly impacts when, where, and how people receive information when information is layered with multiple sources providing independent, dependent, and inter-related information. This impact is evident from a 2014 Council for Research Excellence study that revealed that 16% of prime-time viewing occasions (including news delivery) involved interaction with social media about the television event being watched, which created so-called "socially connected viewing" [33].

In Other Words...Increasing Media Consumption

One can actually have more than 24 hours in a media day. As we increase the number of simultaneous media streams going into the home and increase our multi-tasking behaviors, a lot of content assumes the role of background or secondary content streams. And as we increase our level of multi-tasking, we have to expect that the total hours will grow even as the total number of physical hours a viewer can consume media will remain roughly constant.

~James E. Short, University of Southern California's Marshall School of Business [32]

Interestingly, this increasing consumption of media and data is not necessarily without limits. For example, between 2012 and 2013, the amount of video watched by Americans stayed steady, but the average length of each video shrank from almost 7 minutes to just over 5 minutes [34]. Likewise, a separate British study found that an average person switches between digital devices up to 21 times per hour [35]. This pace of information consumption and processing flow must be considered for disaster communications. Traditional processing of regular information released may not be compatible enough with this shifting and overlapping landscape as people shift from source to source to find, aggregate, and verify information to a level of acceptance they are comfortable with.

This level of acceptance represents another major shift in communications and engagement. Traditionally, trusted sources of information have been presented in single-source formats such as the daily newspaper or the evening newscast. However, with the decentralization and diversification of communication sources present with online and digital formats, it is clear that what the public deems as trustworthy is significantly shifted. This shift is particularly challenging for government representatives in public safety and emergency management, who often face significantly inherent trust issues with public constituents throughout the world [36]. For example, Edelman's Trust Barometer for 2014 measured an increasing trust level for various digital formats including online searching and social media. Specifically, the Edelman assessment found that while traditional media were still considered the most trustworthy (65% of those surveyed), online search engines were very close at 63%. Likewise, hybrid media and social media were not far behind at 53% and 45% of the respondents, respectively [36].

Trust can be defined as the relative state of positive expectations about another entity's motives with respect to situations involving risk [37]. To understand why Internet and digital systems have increased in trustworthiness and credibility, a broader consideration of trust must be considered. Relationships based solely on information exchange are particularly important before, during, and after disasters and exist between individual members of shared communities, those communities and traditional media outlets, and between those communities and their government representatives in preparedness, response, and recovery. Each of these trust relationships is built on the need to be able to substitute for each other, influence one another, and have a positive attitude toward each party involved [37]. Conventional media (e.g., television) has traditionally had difficulty serving as a substitute for those needing information; however, that is one of the few strengths of government representatives. In contrast, social media is one of the first formal communication avenues that can effectively support all three facets of strong trust and dependability when it comes to critical communications before, during, and after an event.

The level of trust for social media can further be evaluated by looking at its ability to give information generators and consumers authority, helpfulness, intimacy, and self-promotion [38]. Specifically, the degree of authority, helpfulness, and intimacy all increase trust while self-promotion is the only undermining factor. Social media has simultaneously served to undermine the authority of traditional media and official government communications by providing information in a more timely and transparent methodology with at least the appearance of minimal self-interest. Likewise, social media has routinely served to be efficient and effective at providing directed and practical help during emergency response and recovery. Emergent volunteerism groups that emerge during events and text-to-donate campaigns are commonplace now after disasters and often facilitate individual, family, and community preparedness that exists outside and without the assistance of formal emergency response structures. Social media is also far more intimate than traditional media or government communication sources. Individual users can directly connect with those impacted, provide support, and see the words and pictures (e.g., disaster selfies) of those impacted without filter or repackaged structure. These direct connections can have a powerful impact on the real and perceived trustworthiness of social media messaging. Lastly, while self-promotion is possible (as some have suggested related to the disaster selfies), the collection of social media users within a given system like Twitter or Facebook often self-regulates these behaviors to eliminate, minimize, or otherwise correct information shared before, during, or after disasters that have a real or perceived slant toward self-promotion rather than for the collective good.

To compensate for this shift toward social media, traditional media outlets have begun to more broadly utilize the concept of narrative in news reporting to tap into the power and purpose of social media. While there are numerous definitions of narrative, it is far more than just telling a story. For example, researchers at the Ohio State University describe narrative as a "multidimensional purposive communication from a teller to an audience" that focuses on narrative as a means to ensure that "we are interested in the ways in which the elements of any narrative (e.g., character, setting, etc.) are shaped in the service of larger ends" [39]. The media creates this narrative by leveraging social media itself and through an increasing focus on well-developed, narrative-driven images and visuals that are often filled with primary and superficial information that would not necessarily be considered newsworthy, but does generate depth to the narrative being built around the incident. These visuals are often referred to as infographics and are relatively easy to digest and often extremely cost effective as they can be repeatedly shared via social media outlets.

The utilization of infographics is an underutilized tool by professional emergency managers, which is understandable considering that many emergency managers lack the skills or resources to develop them in a timely manner. However, some organizations, like the Federal Emergency Management Agency (FEMA), access a larger pool of resources and partnerships with external organizations (e.g., local newspapers) to repurpose or retool infographics that have been developed for impacted local communities. For example, during the 2011 support of the tornado outbreak in Alabama, FEMA partnered with the *Times-Picayune* newspaper in New Orleans to create an infographic about the collection, sorting, and processing of debris materials after a disaster.

In Other Words...The Need for Narrative

The media isn't the only one that needs a narrative...we do. We need to make sense of what's around us, not just the true things that really happened, but the fictional ones that we know didn't. All this myth-making reminds us how strongly wired we are to believe in things that both make sense and feel right. They feel right because of who told us and when. Culture creates reality.

~Seth Godin [40]

The last major impact from this communications transition is the physiological changes that have been identified from the increase and widespread use of social media and other digital communication systems. Specifically, the diversity of digital sources and the depth of information received and disseminated are changing how human beings are able to process information and prioritize life choices. For example, in one study of social media users aged 18–85, resisting the urge to use Facebook and Twitter was harder than resisting smoking, drinking, sex, sleeping, or spending money [41]. Researchers believe that social media utilizes the same brain chemical pathways for dopamine that register positive experiences related to food, sex, and other beneficial activities. Likewise, in a separate study where people were rewarded less money to answer questions about themselves than to answer general questions, 17% to 25% took less money than the possible earnings to talk more about themselves [41]. This engagement seems unique to active engagement of social media systems like Facebook, as some studies have shown a decrease in the positive effects when engagement is passive (e.g., simple Internet searching) [42].

Understanding this layered communication system is critical for emergency managers. Moreover, as the trends show no indication of a reduction of use or application of social media and digital communications, it is a fair assessment to project that this process will continue as traditional systems either give way to newer communication forms or become hybrid offerings of both traditional and digital forms of communication. Emergency management professionals must be intentional in the processes created to accept and process these new information sources because "we're entering a new era…in which…experiences are not centered on physical objects but on the fabric of digital information that surrounds us" [43].

REFERENCES

1. Gladwell, Malcolm. (2013). *David and Goliath: Underdogs, Misfits, and the Art of Battling Giants*. New York: Little, Brown & Company.
2. McDayter, Mark. (2012). "A (Very) Brief History of Printing." Printing Technology in the Handpress Era—The Printer's Devil Project. http://ett.arts.uwo.ca/printersdevil/site/printing/history.html. Accessed April 3, 2014.
3. Curtis, Anthony. (2013). "A Brief History of Journalism in America." University of North Carolina at Pembroke. http://www2.uncp.edu/home/acurtis/Courses/ResourcesForCourses/JournalismHistory.html. Accessed April 3, 2014.
4. "Sixty Years of Newspaper Circulation Trends." (2011). Communications Management, Inc. http://media-cmi.com/downloads/Sixty_Years_Daily_Newspaper_Circulation_Trends_050611.pdf. Accessed April 3, 2014.
5. Sicha, Choire. (2009). "A Graphic History of Newspaper Circulation over the Last Two Decades." The AWL. http://www.theawl.com/2009/10/a-graphic-history-of-newspaper-circulation-over-the-last-two-decades. Accessed April 4, 2014.
6. Boulton, Terynn. (2013). "Who Is Craig from Craigslist?" Today I Found Out. http://www.todayifoundout.com/index.php/2013/09/who-is-craig-from-craigslist/. Accessed April 4, 2014.
7. "Craigslist Statistics." (2014). Statistics Brain. http://www.statisticbrain.com/craigslist-statistics. Accessed April 4, 2014.
8. Rubin, Marina Koestler. (2010). "Radio Activity: The 100th Anniversary of Public Broadcasting." *Smithsonian*. http://www.smithsonianmag.com/history/radio-activity-the-100th-anniversary-of-public-broadcasting-6555594/?no-ist. Accessed April 4, 2014.
9. Santhanam, Laura et al. (2013). "Audio by the Numbers." The State of the News Media 2013. http://stateofthemedia.org/2013/audio-digital-drives-listener-experience/audio-by-the-numbers. Accessed April 5, 2014.
10. Berkowitz, Justin. (2010). "The History of Car Radios." *Car and Driver*. http://www.caranddriver.com/features/the-history-of-car-radios. Accessed April 6, 2014.
11. Ramsey, Mark. (2013). "Check Out the Reach for Streaming Audio." Mark Ramsey Media (MRM). http://www.markramseymedia.com/2013/02/check-out-the-reach-for-streaming-radio/. Accessed April 7, 2014.

12. Israel, David K. (2010). "A Short History of the Emergency Alerting System (EAS)." *Mental Floss.* http://mentalfloss.com/article/25704/short-history-emergency-broadcast-systems. Accessed April 7, 2014.
13. "Emergency Alerting System." (n.d.). Federal Communication Commission (FCC). http://www.fcc.gov/guides/emergency-alert-system-eas. Accessed April 7, 2014.
14. "Wireless Emergency Alerts." (2013). CTIA Wireless Association. http://www.ctia.org/your-wireless-life/consumer-tips/wireless-emergency-alerts. Accessed April 8, 2014.
15. "Wired, Zapped, and Beamed: 1960s through 1980s." (2005). US Federal Communications Commission (FCC). http://transition.fcc.gov/omd/history/tv/1960-1989.html. Accessed April 8, 2014.
16. "Television Watching Statistics." (2013). *Statistics Brain.* http://www.statisticbrain.com/television-watching-statistics/. Accessed April 9, 2014.
17. "Key Indicators in Media and News." (2014). Pew Research Journalism Project. http://www.journalism.org/2014/03/26/state-of-the-news-media-2014-key-indicators-in-media-and-news/. Accessed April 9, 2014.
18. Sedghi, Ami. (2014). "Facebook: 10 Years of Social Networking, in Numbers." *The Guardian.* http://www.theguardian.com/news/datablog/2014/feb/04/facebook-in-numbers-statistics. Accessed April 11, 2014.
19. Crowe, Adam. (2012). *Disasters 2.0: The Application of Social Media in Modern Emergency Management.* Boca Raton, FL: CRC Press.
20. Arthur, Charles. (2008). "How Twitter and Flickr Reported the Mumbai Terrorist Attacks." *The Guardian.* http://www.theguardian.com/technology/2008/nov/27/mumbai-terror-attacks-twitter-flickr. Accessed April 12, 2014.
21. "Citizen Journalism." (n.d). *Mashable.* http://mashable.com/category/citizen-journalism. Accessed April 12, 2014.
22. Sambrook, Richard. (2005). "Citizen Journalism and the BBC." Harvard Nieman's Reports. http://www.nieman.harvard.edu/reports/article/100542/Citizen-Journalism-and-the-BBC.aspx. Accessed April 12, 2014.
23. Stanglin, Doug. (2013). "Student Wrongly Tied to Boston Bombing Found Dead." *USA Today.* http://www.usatoday.com/story/news/2013/04/25/boston-bombing-social-media-student-brown-university-reddit/2112309. Accessed April 13, 2014.
24. Kaufman, Leslie. (2013). "Bombing Trips up Reddit in Its Turn in the Spotlight." *The New York Times.* http://www.nytimes.com/2013/04/29/business/media/bombings-trip-up-reddit-in-its-turn-in-spotlight.html?_r=0. Accessed April 13, 2014.
25. Busher, Monika et al. (2012). "Altruistic, Augmented, Agile: Public Crisis Response." BRIDGE Project. http://www.bridgeproject.eu/en/bridge-results/publications/altruistic-augmented-agile. Accessed April 15, 2014.
26. Szalavitz, Maia. (2012). "Is Human Nature Fundamentally Selfish or Altruistic?" *Time.* http://healthland.time.com/2012/10/08/is-human-nature-fundamentally-selfish-or-altruistic. Accessed April 15, 2014.

27. Chamorro-Premuzic, Tomas. (2014). "Sharing the (Self) Love: The Rise of the Selfie and Digital Narcissism." *The Guardian.* http://www.theguardian.com/media-network/media-network-blog/2014/mar/13/selfie-social-media-love-digital-narcassism. Accessed April 16, 2014.
28. Stewart, Dodai. (2014). "The Rise of Disaster Survivor Selfie." *Jezebel.* http://jezebel.com/the-rise-of-the-disaster-survivor-selfie-1562194440. Accessed April 16, 2014.
29. Morrison, Sarah. (2014). "Here Is the Inevitable Plane Crash Selfie." *The Wire.* http://www.thewire.com/culture/2014/03/here-inevitable-plane-crash-selfie/359162. Accessed April 19, 2014.
30. Robbins, Mel. (2014). "Opinion: School Stabbing Victim's Selfie Was Just Fine." *CNN Opinion.* http://www.cnn.com/2014/04/09/opinion/robbins-selfie-school-stabbing/. Accessed April 19, 2014.
31. Young, Brenda. (2009). "Human Capacity for Information Is Massive, but Not Infinite." *Tech News World.* http://www.technewsworld.com/story/68865.html. Accessed April 19, 2014.
32. "US Media Consumption Rises to 15.5 Hours a Day Per Person—by 2015." Science X Network. http://phys.org/news/2013-11-media-consumption-hours-dayper-personby.html. Accessed April 19, 2014.
33. McAdams, Deborah D. (2014). "Research: One-Fifth Use Social Media for TV." TV Technology. http://www.tvtechnology.com/news/0086/research-one-fifth-use-social-media-for-tv/269873. Accessed April 19, 2014.
34. Greenfield, Rebecca. (2013). "The Internet's Attention Span for Videos Is Quickly Shrinking." *The Wire.* http://www.thewire.com/technology/2013/08/internets-attention-span-video-quickly-shrinking/68114. Accessed April 19, 2014.
35. Smith, Jennifer. (2014). "Proof of Our Shrinking Attention Span." *Daily Mail.* http://www.dailymail.co.uk/news/article-2534163/Proof-shrinking-attention-span-Average-person-switches-devices-21-times-HOUR.html. Accessed April 20, 2014.
36. "Edelman Trust Barometer—Executive Summary." (2014). http://www.edelman.com/insights/intellectual-property/2014-edelman-trust-barometer/about-trust/executive-summary. Accessed April 20, 2014.
37. Lewicki, Roy J. and Bunker, Barbara B. (n.d). "Trust in Relationships." Diamond Management Consultants. http://cms.nortia.org/Org/Org134/Groups/Resource%20Centre/Diamond%20Resources/comp10_TrustInRelationships.pdf. Accessed April 21, 2014.
38. Rayson, Steve. (2013). "The New Formula for Social Media Trust." *Social Media Today.* http://socialmediatoday.com/steve-rayson/1870091/new-formula-social-media-trust. Accessed April 21, 2014.
39. Herman, David. (2012). *Narrative Theory: Core Concepts and Critical Debates.* Columbus, OH: The Ohio State University Press. https://ohiostatepress.org/Books/Book%20PDFs/Herman%20Narrative.pdf. Accessed April 22, 2014.
40. Godin, Seth. (2013). "The Media Needs a Narrative." Seth's Blog. http://sethgodin.typepad.com/seths_blog/2013/11/the-media-needs-a-narrative.html. Accessed April 22, 2014.

41. Karbach, Med. (2014). "The Impact of Social Media on Our Brain." *Educational Technology and Mobile Learning.* http://www.educatorstechnology.com/2013/03/the-impact-of-social-media-on-our-brain.html. Accessed April 23, 2014.
42. Konnikova, Maria. (2013). "How Facebook Makes Us Unhappy." *The New Yorker.* http://www.newyorker.com/online/blogs/elements/2013/09/the-real-reason-facebook-makes-us-unhappy.html. Accessed April 24, 2014.
43. Dadich, Scott. (2013). "The Age of Invisible Design." *Wired*. September 2013.

3

Data Mining and Predictive Behavior

> That's the next great challenge for design: weaving the threats of technology, information, and access seamlessly and elegantly into our everyday lives. When a social network automatically checks us into a location, or cashiers can suggest new products based on purchase history, or our connected TV calls up our favorite shows when we walk into the living room...it may seem like magic.
>
> ~Scott Dadich, "The Age of Invisible Design" [1]

INTRODUCTION

In a book such as this that evaluates the effects of current trends on future actions and behaviors, the concept of prediction is at the core of each and every chapter. However, for this chapter, the focus and consideration of prediction shift away from macroconsiderations of prediction and focus more on the microcharacteristics of emerging technologies able to predict the behavioral choices of their users. In turn, the technology then responds in ways to anticipate those choices in advance and often without the awareness of the technological user. This type of technological functionality provides an extremely personalized experience with the information and providing technology, which ultimately is beginning to create similar expectations in broader applications—both within and external to the technology.

Before understanding the technological capabilities of predictive behavior, the concept of predictive modeling must first be considered. Predictive behavior modeling is the science of the application of mathematical and statistical techniques to historical and transactional data of customers to predict their future behavior [2]. By modeling future decisions, retention experts can make operational and resource allocation decisions that are more efficient and ultimately more effective than simply using historical examples to try to directly correlate into decision making. Put another way, IBM describes predictive modeling as having "the power to discover hidden relationships in…volumes of structured and unstructured data and us[ing] those insights to confidently predict the outcome of future events and interactions" [3].

From a marketing and business perspective, the benefits of predictive modeling and analysis are widespread. Specifically, this type of modeling typically has a positive impact on the optimization of existing processes, clarification of customer behavior, identification of unexpected opportunities, and anticipation of problems before they are impactful [4]. Likewise, it has routinely been shown to improve an organization's ability to up-sell products, cross-sell services, and improve customer retention campaigns as well as increase the relevance of the organization's communication processes with clientele [3]. Interestingly, even with the wide spectrum of potential positive impacts only 40% of organizations have partially or fully implemented predictive modeling strategies [4] (see Figure 3.1).

This type of modeling would be extremely beneficial to emergency management and public safety personnel. Given that these organizations (with rare exceptions) are hindered by limited resources, there is constantly a need to assess projects and programs to ensure that limited resources are prioritized effectively. Unfortunately, most emergency management organizations simply make planning, organizational, and resource management decisions based on a direct correlation to historical events and actions. The flaw with this technique is that it lacks the forecasting presence and predictive nature of discussed modeling. For example, public safety decisions made strictly on historical patterns would neglect the changing public expectations and communication strategies already discussed in Chapters 1 and 2, which clearly are altering future choices people make.

While there is no one single way to facilitate predictive modeling, there are some clear trends toward a few common approaches to this process. In most cases, the modeling methods require the quantification of risk based on all data, metrics, and measurements which can be collected about an individual user. This collection is often referred to as data mining,

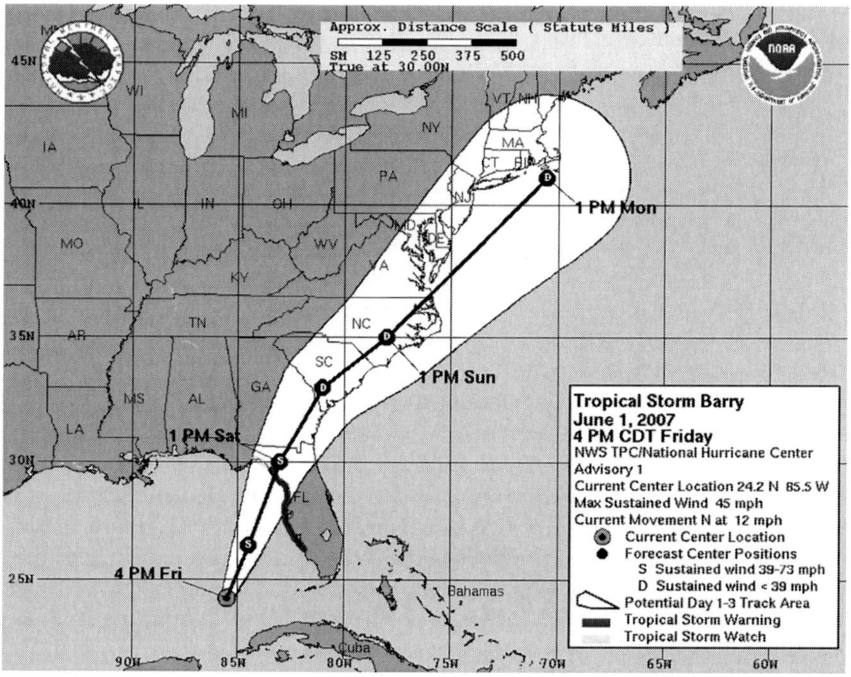

Figure 3.1 Predictive and forecasting technologies such as storm prediction tracks are often utilized, but other predictive technologies lack a similar level of usage. (Source: US National Weather Service.)

which has grown increasingly controversial in certain sectors like homeland security and national intelligence, but represents an extremely valuable and likely future application in all sectors within emergency management and public safety.

While data mining is highly complex and something that will most likely require professional support if fully leveraged in emergency management, the basics must be understood to begin to consider how it can be properly deployed now and in the future. In the simplest form, data mining is "the process of finding logical patterns in data and according order and meaning to various sets of seemingly random data" [5]. Other sources refer to these logical patterns as "knowledge," which is predictive modeling or the ability to take such knowledge and make accurate and discrete projections of future decisions and choices that may affect operational effectiveness or output [6].

The concept of data mining was first introduced in the 1990s, but has a long history tied back to classical statistics, artificial intelligence, and machine learning. The fundamental elements of data mining and ultimately predictive forecasting are a statistical evaluation including regression analysis, standard distribution, standard deviation, standard variance, discriminant analysis, cluster analysis, and confidence analysis [7]. However, this chapter will not consider these components in greater detail as they are unnecessary to understand at this interval. Instead, the possible applications of predictive modeling within disaster management operations will be the primary focus.

In a practical business sense, predictive modeling is widespread, but most often unknown to the end users. As a primary example, the vast majority of free online services, like e-mail or web browsing, are available at no cost due to the embedded information being gathered on the user's web browsing patterns, demographic data, purchasing choices, and search parameters (see Figure 3.2). Even though often unrecognized by the user, it is most evident in the sponsored advertisements, banners, pop-up messages, and other marketing tools that often appear immediately in response to browser actions. For example, if an Internet user utilizes a web browser to search for shoes, the next visited web page with embedded advertisements will promote available shoes, stores, deals, or other incentives to address the previously searched-for parameters. It is this

Figure 3.2 Data mining of free online services gathers browsing patterns, demographic data, and other information about the specific users. (Source: FEMA/Marvin Nauman.)

automated and seemingly intuitive digital relationship that is often seen as predictive or "smart" in its application of a wide spectrum of data.

In Other Words...The Power of Data Mining

Data embodies a priceless collection of experience from which to learn. Every medical procedure, credit application, Facebook post, movie recommendation...and purchase of any kind—each positive or negative outcome, each successful or failed event or transaction—is encoded as data and warehoused. As data piles up, we have ourselves a genuine gold rush. But data isn't the gold—data in its raw form is boring crud. The gold is what's discovered therein. With the new knowledge gained, prediction is possible.

~Eric Seigel [8]

This phenomenon is also present within the digital tools utilized by many millions of people worldwide. For example, with more than 425 million active users, Google's Gmail system is widely utilized as a free e-mail service for people to send and receive e-mails with embedded and attached data. Because of the vast amount of information exchanged, Google aggregates the data and can provide embedded advertisements in response to words used within e-mails as a predictive modeling technology. For example, if an e-mail message contained references to a Native American name, Google might generate an advertisement relating to visiting Native American historical sites. [5,6].

While people certainly enjoy the possibilities of predictive modeling and behavior, there are concerns related to the process of the data mining necessary to achieve such activities. Specifically, the most significant concern related to data mining is the impact to both individual and collective privacy of the users of digital systems. This is particularly concerning if specific persons' names, aliases, social security numbers, e-mail addresses, bank account numbers, and other personal information are revealed via data mining; this leaves the possibility that private marketers, commercial companies, and/or government might utilize this information for unethical or nefarious purposes [9]. These types of ethical challenges related to the protection and preservation of an individual's personal data are the most significant potential hindrance for the future of data mining with particular concern related to how government intelligence organizations have utilized and will utilize such data.

USING DATA MINING FOR INTELLIGENCE GATHERING

The US federal government utilizes data mining for a wide variety of programs. For example, according to one GAO survey, nearly 200 data mining efforts are underway by federal government agencies that seek out and leverage both personal and digital behavior data. Of those identified, nearly 35% were directed at service improvement or performance programming, 12% targeted fraud and waste detection, 11% analyzed scientific and research information, and an additional 8% were used to detect criminal activities and behaviors [9]. It is this last characteristic that has the widest implications on public safety, homeland security, intelligence, and emergency management functions now and into the future.

For example, the Markle Foundation's Taskforce on National Security in the Information Age stated that "information analysis is the brain of homeland security... [if] used well, it can guide strategic, timely moves throughout...the world... [and if] done properly, even armies of guards... will be useless" [10]. Specifically, collecting and analyzing all available data under predictive modeling is far more reliable, accurate, and valuable than simple linear associations in the field of homeland security. For example, identifying a connection between individual suicide bombers and religious extremism adds little value to an organization's ability to combat terrorism, but making connective predictions would increase the likelihood that information about the when, where, and how of planned terrorist events may occur [10]. While nearly all intelligence organizations handle and process data mining at some level, the US National Security Agency (NSA) is one of the most well-known and notorious agencies utilizing these capabilities (see Figure 3.3).

The criticality of data mining and predictive behaviors for national intelligence systems increased in its importance after the terrorist attacks on September 11, 2001. This change in relevance was absolutely critical given the highly decentralized approach of the terrorists responsible for the attacks. Specifically, terrorist cells were both independent and connected, as well as spread throughout the world with minimal use of complex systems and often in the process of long-standing preparations for the assigned tasks. Additionally, the information systems and data

Figure 3.3 The US National Security Agency (NSA), as well as numerous other governmental agencies, has been given the authority to use data mining to gather and leverage user information for various reasons. (Source: FEMA/Bradley Carroll.)

exchanged in these decentralized cells often leveraged the full complexity of digital technology. For example, anonymous e-mail accounts or pay-as-you-go (also known as "burner") mobile phones were leveraged to send, receive, and ultimately act on such information. Consequently, without data mining, it became abundantly clear that it was nearly impossible to detect, collect, analyze, and ultimately decipher nefarious information from terrorists when it was simply buried in nearly 2.5 quintillion bytes of other data exchanged every day [11].

Consequently, intelligence organizations like the NSA quickly moved to expand data mining processes to help more clearly establish connections between people and information. These systems quickly began to put filters or aggregators on information to—in essence—force the astronomical amount of data into measurable "pipe" information where the patterns and connectivity could begin to be seen. In other words, as the data-mining saying goes: "To find a needle in a haystack, you need to first build a haystack" [12]. The building of the so-called digital haystack most frequently relies on extra data that are added as a "tag" onto the primary data. These tags are called metadata and help allow commercial companies and ultimately intelligence organizations to search and filter in order to pull desired data or behavior patterns. Interestingly, metadata tags have a murky legal status when compared to traditional or direct communication methods. For example, the NSA (or other government intelligence-gathering organizations) cannot examine the communications of a US citizen or resident alien, but there are no such limitations on an individual's metadata [11].

The use of metadata to collect information and predict future behavior is ostensibly to prevent future terrorism or other acts that impact public safety. Unfortunately, this process is intentionally wrapped in subterfuge to prevent the public from knowing the process and ultimately finding a way to circumvent the detection. For example, China's Public Security Bureau adds additional goals of public opposition to government leaders and minimizing public opposition to government decisions and operations [12]. Under any approach, the public provides information to the proverbial equation, but has no engagement and little choice in the process. The ubiquitous nature of digital systems and corresponding devices makes avoidance of "digital haystacks" extremely difficult for the average user. Consequently, when a former American government intelligence analyst named Edward Snowden released classified information about data-mining systems in June 2013, the world took notice and began to question the processes and appropriateness of these systems.

Specifically, Snowden took nearly 200,000 classified documents that he had access to and fled to Hong Kong and later Russia. Snowden quickly released the documents to the traditional media in various international markets, who began to identify and analyze information about the inner workings of the US intelligence community. For example, traditional media outlets quickly identified that the NSA collected records of every American phone call under a call log metadata program. Moreover, the *Washington Post* revealed that the NSA infiltrated the cloud-based services of Google and Yahoo to collect the data of America's digital profiles and activities. Other European media outlets determined from Snowden's released documents that the United States collected data on allies, including Germany, France, and Italy [13].

Additionally, a Snowden-leaked document revealed an NSA surveillance program called US-984XN, which was more commonly known by its code name, PRISM, which was established by President George W. Bush in 2007 as a covert program for warrantless domestic surveillance (see Figure 3.4). PRISM involved the collection of digital photos, online storage data, file transfers, e-mails, chat room logs, videos, and video conferencing data from nine major American Internet companies [11].

Figure 3.4 US President George W. Bush initiated a covert digital surveillance program in 2007 which was further developed under President Obama and later revealed by NSA whistleblower Edward Snowden. (Source: White House/Eric Draper.)

Specifically, through PRISM, the NSA was able to extract information directly from the servers of companies like Microsoft, Yahoo, Google, Facebook, AOL, Skype, YouTube, and Apple [14]. Likewise, it was quickly confirmed that the Government Communications Headquarters (GCHQ; Great Britain's equivalent to the NSA) had also been secretly data mining information from the same Internet companies through an operation set up by the NSA. Specifically, this NSA-generated system would allow GCHQ to circumvent the formal legal process required in Great Britain to collect personal information included in e-mails, photos, and photos from Internet companies based outside the United States [14].

The challenge is that these systems are not foolproof. The effectiveness and appropriateness of these systems must be taken in balance when considered against the impacts to privacy and questions related to legal and ethical application of data mining. Well before Snowden released the classified documents, many information security experts were already questioning the effectiveness of data mining on the identification of terrorists and the prevention of future acts of violence. For example, in a 2007 sworn testimony to the Committee on the Judiciary of the United States Senate, security experts stated that "with a relatively small number of [terrorist] attempts every year and only one or two major terrorist incidents every few years—each one distinct in terms of planning and execution—there are no meaningful patterns that show what behavior indicates planning or preparedness for terrorism" [15]. Likewise, routine and acceptable behavior of the general population can lead to its misidentification as questionable behavior. For example, after the September 11 terrorist attacks, many government leaders and emergency response organizations questioned (and even arrested) individuals for taking pictures of bridges, monuments, and buildings (see Figure 3.5). While this behavior may be used for nefarious purposes, it is much more likely a pattern of innocent behavior related to tourism or hobby photography, for example [15].

Regardless of these limitations, concerns, and political and public black eyes, intelligence organizations are continuing to move forward with the dedication of resources, personnel, and procedures to the use of data mining for homeland security and terrorism prevention. For example, the NSA is building a $1.7 billion facility in Utah that will facilitate the storage and processing of data-mining information and related classified information. This facility will be the largest data storage center in the United States and will constantly use 65 megawatts of electricity, which is enough to power 33,000 houses. The NSA is maintaining a high level of secrecy related to the facility and will not reveal any specifics about the operations or structure of the location [16]. This facility was initially

Figure 3.5 After 9/11, there was a significant concern about individuals who took pictures of bridges, monuments, buildings, and other pieces of critical infrastructure. (Source: US Navy/Sgt. Andy Dunaway.)

welcomed by politicians in Utah with a promise that activities would be "conducted according to constitutional law," but many national leaders in as many as 10 states have introduced legislation to limit or withdraw funding to support this effort [16]. The challenge of the commitment and possible uses of data mining by the NSA most likely will continue to run contradictory to public and political pressure until a balance of privacy and protection can be found.

PUBLIC SAFETY USES OF DATA MINING AND PREDICTIVE MODELING

National homeland security and intelligence organizations are not the only groups using data mining and predictive modeling. Law enforcement entities at the local, state, and federal levels of government are beginning to utilize and consider future applications of these technological capabilities. Much like the intelligence communities already mentioned, law enforcement organizations are burdened by a similar amount of data that can quickly overwhelm professional analysts without data mining and predictive modeling possibilities.

The use of data mining and predictive modeling in law enforcement is a relatively new concept. Back in 2009, seven American police organizations received planning grants through the National Institution of Justice's (NIJ) competitive solicitation process to consider how digital information could be collected and analyzed to prevent and/or reduce crime in various communities. These organizations included police departments from large metropolitan areas including Los Angeles, Boston, Chicago, New York, and Washington, DC [17]. These entities and many since have begun to utilize predictive policing in four major areas: predicting crimes, predicting offenders, predicting the identity of perpetrators, and predicting victims of crimes. Predicting crimes is the broadest of these categories and focuses on the forecasting of places and times that have an increased risk of crime. Similarly, other models attempted to predict the identity of individuals at increased risk of committing crime in the future. The identity of perpetrators can also be projected against certain profiles that can more accurately predict likely offenders with specific past crimes. The last and perhaps most interesting classification model being utilized is related to the prediction of victims of crimes. In all four cases, leveraging these types of models could potentially have a significant impact on public safety and clearly is becoming commonplace and will likely become a best practice across all types of organizations as resources and comfort with technology become more available [18].

Much like the intelligence community utilization of predictive modeling, there are limitations related to how law enforcement can effectively use these systems. For example, some people have argued that effective predictive crime simply displaces the crime to another geographic area or jurisdiction. However, these types of challenges are often countered in the algorithms utilized in the data processing and/or create a halo effective by having positive, yet unintentional, impacts on other areas [19]. Additional limitations include an over-reliance on predictions and/or erroneous data. For example, some law enforcement agencies lack effective strategies to transition from the predictions to tactical application to actually stop or reduce crime in the area. Likewise, if erroneous data or related assumptions are applied, the predictions applied to the collected data can lead to misapplied resources and ultimately reduce the impact of reducing crime or the effectiveness of overall public safety initiatives [18].

PERSONAL PREDICTIVE BEHAVIOR

Predictive behavior technology is not limited to professional and widespread uses in public safety and emergency management. Many newer digital systems and mobile technologies such as cell phones and tablet computers have embeddable technology (or applications) that serve as so-called virtual personal assistants. Examples of these virtual personal assistants include Siri by Apple, Google Now, Mynd, and Cortana by Microsoft. In most cases, these software programs are voice activated and respond to requests made by the user (e.g., location of building, message generator, etc.). While impressive in their own right, these systems are also ultimately designed to recognize patterns and design in the behaviors of the device user.

These prediction-based virtual personal assistants have and will continue to grow in importance as various forms of emerging and disruptive technologies integrate with various mobile devices (see Chapter 4). Consequently, companies like Apple, Google, and Microsoft will continue to invest in the capability and reliability of these systems. As they improve the capabilities and streamline these potential integrations, the possibilities for these systems for respective public safety and emergency management personnel and organizations are vast. While operational applications are still limited given the minimal amount of data available in most communities (i.e., the lack of emergencies and disasters personally handled by staff), the day-to-day applications are far more promising. Using predictive technologies embedded in mobile phones or tablets to cut down on the device recall for work schedules, operational conditions (e.g., weather), task reminders, and automated messages will ultimately improve the efficiency and effectiveness of individual practitioners.

In Other Words...Expectations of Predictive Technologies

How could there be anything wrong with this conventional design paradigm? Functionality? Check. Content? Check. Customer personas? Ah—herein lies the problem. These aggregate representations of your customers can prove valuable when designing apps and are supposedly the state of the art when it comes to customer experience and app design, but personas are blind to the needs of the individual user. Personas were fine in 1999 and maybe even in 2009—but no longer, because we live in a world of 7 billion "me's." Customers increasingly expect and deserve to have a personal relationship with the hundreds of brands in their lives. Companies that increasingly ratchet up individual experiences will succeed. Those that don't will increasingly become strangers to their customers.

~Mike Gualtieri, Forrester Blogs [20]

However, like the data mining and broad-scale predictive behavior mentioned earlier in the chapter, the personal uses of these technologies also have limitations and ethical concerns regarding deployment. For example, whether the mobile device is on-demand (e.g., Siri) or in a state of constantly listening (e.g., Google Now), the device is constantly monitoring, collecting, and processing personal information ranging from location to online searching patterns. This type of technology can create "Big Brother" ethical and privacy concerns [20]. Each software developer and device designer will have to address this privacy issue in ways that are consistent with the product and brand management (similarly to the large-scale balance mentioned earlier). For example, Microsoft's Cortana project has a built-in dashboard to allow users to see exactly what information the software is tracking and the types of data being saved [21].

SHIFTING TOWARD SMART BEHAVIORS

Businesses are utilizing data mining and related mathematical algorithms to collect and aggregate customer (or potential customer) activities into predictable or measurable behaviors. Because customers do not directly engage in these activities and frankly are often unaware they are occurring, there is a tremendous spectrum of possible uses and impacts. These measurable behaviors are being leveraged to create organizational opportunities that can increase profit, improve client experience, and strengthen customer service. Likewise, as will be discussed in this section, there are also significant potential implications on how this type of behavior can potentially be used to influence the behavior of individuals before, during, and after disasters.

For example, the Target organization assigns every customer an identification number that is tied to credit card numbers, purchasing history, e-mail addresses, personal mailing address, and other personal data. Once these are collected Target often looks to send targeted coupons and other marketing material via mail or e-mail. For example, Target sends coupons to women who appear (via predictive behavior) to be pregnant or trying to conceive. After analyzing years of purchasing data of pregnant women, Target administrators determined that women on the baby registry purchased larger quantities of vitamin supplements (e.g., calcium) in their first trimester, unscented lotion at the beginning of their second trimester, and cotton balls and hand sanitizers just prior to their delivery date (see Figure 3.6). Target utilized these product-purchasing patterns as well as those for 22 other products to create a "pregnancy prediction" score which was then used to trigger targeted coupons and other promotional materials. In the end, Target's revenue from pregnant (or trying to get pregnant) women grew from $44 billion in 2002 to $67 billion in 2010 due to the data-mined process of "heightened focus on items and categories that appeal to specific guest segments such as mom and baby" [22].

Online retailer Amazon is also attempting to potentially leverage customer behavior patterns to reduce shipping time for their customers. According to an Amazon patent filed in 2014, the company is attempting to utilize previous orders, system searches, wish lists, and cursor hovering time to create "anticipatory shipping" where the products bought by the customer start traveling to the nearest shipping hub before a purchase button has been clicked [23]. While Amazon's predictive shipping model is only a theory at the time of print, it is a strong indicator that private

Figure 3.6 Target infamously utilized data mining to create predictive modeling around how pregnant women would buy goods and products. (Source: FEMA/ Liz Roll.)

companies have sought and will continue to seek out ways to predict behavior to improve operational processes and ultimately the financial impact to the organization.

Although still not commonplace, these sorts of predictive behavior models are being directly utilized to improve community and individual health, safety, and emergency preparedness. For example, in 2014, Google announced the $3.2 billion acquisition of a company called NEST Labs, which is a maker of home automation equipment such as thermostats and smoke/carbon monoxide detectors. Since 2011, NEST has made products that are "smart" and ultimately learn from the behaviors and choices of the end user. For example, the thermostat is programmable (like any other thermostat), but the NEST products learn desired temperatures and raise or lower the temperatures based on occupancy patterns (e.g., workday vs. weekend) in the area, which ultimately improves energy efficiency and lowers costs [24,25]. Likewise, the "smart" smoke and carbon monoxide detector responds to safety risks with a human voice and a "friendly heads-up" that includes information about in which room the smoke or carbon monoxide is located [24]. The devices also constantly monitor battery life and send messages to a smartphone app when they need to be replaced, which is both more efficient and less arbitrary and impactful than currently recommended preparedness strategies [26]. While this type of predictive modeling system is not without integration challenges, it yet again shows a shift toward processes that allow for and support the use of past behaviors to predict future choices and activities.

LEVERAGING EMOTIONAL CONNECTIONS

One of the strongest possible benefits of using data mining and predictive modeling for business and commercial practices is the possibility of eliminating hurdles for the consumer and thus providing a completely positive experience with the brand or product line. These types of positive experiences create emotional connections (positive or negative) and ultimately can further be leveraged for long-term brand connectivity and loyalty. Specifically, organizations will utilize emotional branding to appeal to a customer's emotional state, ego, desires, and needs [27]. When these emotion-based characteristics are leveraged, consumers often engage in a self-fulfilling prophecy of consumer engagement. Brands like Nike and Timberland create emotional stories and narratives (see Chapter 2) around consumer experiences by creating heroes or lifestyles around those individuals to whom, ultimately, the consumer is emotionally attached or wishes to emulate [27].

While some consumer groups have objected to all forms of emotional branding as manipulative of human emotions, it is important for public safety and emergency management professionals to understand the power of putting local citizens or a broader constituency in positive emotional positions to make more effective and predictable decisions before, during, and after a disaster. This emotional attachment is why many emergencies or disasters are defined by an emotional photo, video, or narrative of people uniquely impacted by the event. This emotional connection also drives fund-raising initiatives, donations management, and volunteerism rates during certain events. If this concept is more broadly understood, it may be possible to leverage it for expected behaviors and actions during emergencies or disaster recovery activities.

REFERENCES

1. Dadich, Scott. (2013). "The Age of Invisible Design." *Wired*. September/October 2013.
2. "Predictive Behavior Modeling." (2013). Optimove Learning Center. http://www.optimove.com/learning-center/predictive-behavior-modeling. Accessed April 25, 2014.
3. "Predictive Modeling." (n.d.). IBM. http://www-01.ibm.com/software/analytics/solutions/predictive-modeling. Accessed April 25, 2014.
4. Eckersen, Wayne W. (2007). "Predictive Analytics: Extending the Value of Your Data Warehousing Investment." The Data Warehousing Institute. http://www.sas.com/events/cm/174390/assets/102892_0107.pdf. Accessed April 25, 2014.
5. Saitta, Sandro. (2010). "What Is Data Mining?—Explaining It to the Layman." *Data Mining Research*. http://www.dataminingblog.com/guest-post-what-is-data-mining-%E2%80%93-explaining-it-to-the-layman/. Accessed April 27, 2014.
6. Ludwig, Sean. (2012). "Gmail Finally Blows Past Hotmail to Become the World's Largest Email Service." *Venture Beat News*. http://venturebeat.com/2012/06/28/gmail-hotmail-yahoo-email-users. Accessed April 29, 2014.
7. "Data Mining." (n.d.). University of North Carolina. http://www.unc.edu/~xluan/258/datamining.html. Accessed April 27, 2014.
8. Wolverton, Joe. (2013). "Predictive Analytics: The Power to Predict Human Behavior." *The New American*. http://www.thenewamerican.com/reviews/books/item/15660-predictive-analytics-the-power-to-predict-human-behavior. Accessed April 26, 2014.
9. "Think Before You Dig: Privacy Implications of Data Mining and Aggregation." (2004). NASCIO. http://www.nascio.org/publications/documents/nascio-datamining.pdf. Accessed April 30, 2014.
10. McCue, Colleen. (2005). "Data Mining and Predictive Analysis: Battlespace Awareness for the War on Terror." *Defense Intelligence Journal*. 13-1&2. http://innovative-analytics.com/docs/BattlespaceAwarenessDIJ.pdf. Accessed April 26, 2014.
11. Pappalardo, Joe. (2013). "NSA Data Mining: How It Works." *Popular Mechanics*. http://www.popularmechanics.com/technology/military/news/nsa-data-mining-how-it-works-15910146. Accessed May 1, 2014.
12. Harrison, Mark. (2013). "Needles in the Mega-Haystack: NSA Versus KGB." Warwick Centre on Competitive Advantage in the Global Economy. http://blogs.warwick.ac.uk/markharrison/entry/needles_in_the. Accessed May 1, 2014.
13. Kelley, Michael and Nudelman, Mike. (2013). "The Snowden Saga: Here's Everything We Know about the NSA's Nightmare Leak." *Business Insider*. http://www.businessinsider.com/everything-we-know-about-snowden-leaks-2013-11. Accessed May 1, 2014.

14. Gellman, Barton and Poitras, Laura. (2013). "U.S., British Intelligence Mining Data from Nine U.S. Internet Companies in Broad Secret Program." *The Washington Post*. http://www.washingtonpost.com/investigations/us-intelligence-mining-data-from-nine-us-internet-companies-in-broad-secret-program/2013/06/06/3a0c0da8-cebf-11e2-8845-d970ccb04497_story.html. Accessed April 1, 2014.
15. Harper, Jim. (2007). "Balancing Privacy and Security: The Privacy Implications of Government Data Mining Programs." Cato Institute. http://www.cato.org/publications/congressional-testimony/balancing-privacy-security-privacy-implications-government-data-mining-programs. Accessed May 2, 2014.
16. "Electrical Problems Put Damper on NSA's Secretive New $1.7 Billion Data Center in Utah." (2013). *NY Daily News*. http://www.nydailynews.com/news/national/nsa-center-utah-plagued-eletrical-issues-article-1.1479845. Accessed May 2, 2014.
17. "Predictive Policing Symposium: Discussion on the Predictive Policing Demonstration Projects and Evaluations." (2012). National Institute of Justice (NIJ). http://www.nij.gov/topics/law-enforcement/strategies/predictive-policing/symposium/Pages/discussion-demonstrations.aspx. Accessed May 2, 2014.
18. Perry, Walter L. (2013). "Predictive Policing: The Role of Crime Forecasting in Law Enforcement Operations." RAND Corporation. https://www.ncjrs.gov/pdffiles1/nij/grants/243830.pdf. Accessed May 2, 2014.
19. Groff, Elizabeth R. and Lavigne, Nancy G. (n.d). "Forecasting the Future of Predictive Crime Mapping." *Crime Prevention Studies* 13:29–57. http://www.popcenter.org/library/crimeprevention/volume_13/03-groff.pdf. Accessed May 2, 2014.
20. Gualtieri, Mike. (2013). "Predictive Apps Are the Next Big Thing in App Development." Forrester Blogs. http://blogs.forrester.com/mike_gualtieri/13-08-28-predictive_apps_are_the_next_big_thing_in_app_development. Accessed May 3, 2014.
21. Covert, Adrian. (2014). "Google Now and Cortana Are the Future, Not Siri." *CNN Money*. http://money.cnn.com/2014/04/28/technology/innovation/google-now-cortana-siri. Accessed May 3, 2014.
22. Hill, Kashmir. (2012). "How Target Figured Out a Teen Girl Was Pregnant Before Her Father Did." *Forbes*. http://www.forbes.com/sites/kashmirhill/2012/02/16/how-target-figured-out-a-teen-girl-was-pregnant-before-her-father-did. Accessed May 6, 2014.
23. Sorokanich, Robert. (2014). "Amazon Might Try Shipping Things Before You Even Buy Them." Gizmodo. http://gizmodo.com/amazon-might-try-shipping-things-out-before-you-even-bu-1503661403?rev=1389991970&utm_campaign=socialflow_gizmodo_twitter&utm_source=gizmodo_twitter&utm_medium=socialflow. Accessed May 6, 2014.

24. Curtis, Sophie. (2014). "What Is NEST and Why Has Google Bought It?" *The Telegram*. http://www.telegraph.co.uk/technology/google/10570414/What-is-Nest-and-why-has-Google-bought-it.html. Accessed May 6, 2014.
25. Turrentine, Lindsey. (2014). "Second-Gen NEST Zeroes in on Perfection." CNET. http://www.cnet.com/products/nest-learning-thermostat. Accessed May 6, 2014.
26. "Life with NEST." (n.d.) NEST Labs. https://nest.com/smoke-co-alarm/life-with-nest-protect. Accessed May 6, 2014.
27. Barakat, Christine. (2014). "Emotional Branding and the Emotionally Intelligent Consumer." Social Times. http://socialtimes.com/emotional-branding-emotionally-intelligent-consumer_b140449. Accessed May 7, 2014.

4

Emerging and Disruptive Technologies

We tend to overestimate the effect of technology in the short run and underestimate the effect in the long run.

~Roy Amara [1]

"It's all about the T word—Trust."

~John Hanson, Toyota National Manager [2]

HISTORICAL DISRUPTION OF TECHNOLOGY

History is littered with equipment, devices, systems, and resources that once represented the peak of technological implementation. Conversely, entire revolutions in communication and manufacturing have also ridden the wave of new, innovative, and disruptive technologies that were unique and necessary at that given time of history. However, all technology that emerges and ultimately impacts the society and culture which it surrounds is dependent on the context of social, economic, cultural, and demographic characteristics of the historical community for which it first arose and the ongoing acceptance of the technology.

For example, when Alexander Graham Bell invented the telephone in the late nineteenth century it brought about a revolution in person-to-person communication (see Figure 4.1). This revolution continued

Figure 4.1 In the late nineteenth century, Alexander Graham Bell invented the telephone, which started a revolution in person-to-person communication. (Source: Library of Congress.)

into much of the twentieth century with phones shifting from a luxury item to a phone present in most homes to multiple phones throughout every home in developed countries. However, by the end of the twentieth century the impact of land-based telephone systems was no longer revolutionary. In fact, in many ways the "death march" of the landline phone has been progressing for much of the last decade. For example, a 2013 *Wall Street Journal* report stated that just shy of 71% of households in the United States had landline phones, which was down from 96% at the turn of the century. But perhaps more impactful to this previously disruptive technology is the fact that for households under the age of 30 only one-third of those surveyed continued to maintain a landline phone [3].

Clearly not every technology has the historical impact of phones or the longevity of disruptive and revolutionary behavior. However, understanding the technologies that are currently emerging and/or beginning to be disruptive to culture and community will help emergency management and public safety personnel better plan and prepare for their impacts on the community. This includes the need to consider the impact of these emerging technologies on risk (both increasing and decreasing) within communities and to consider whether they can be leveraged to better forecast how individuals will respond before, during, and after disaster events. Consequently, this chapter will review technologies that are emerging across the world that have risen or may rise to the disruptive and impactful nature of some of history's most significant technological developments.

One way to evaluate how emerging and disruptive technologies impact society is by considering where a given technology falls on a hype cycle. The hype cycle concept proposes a comparison of public expectations related to technology over a given period of time. This graphical stepwise process starts with a technological trigger which is immediately followed by a peak of higher expectations, which is quickly followed by a trough of lowered expectations. The initial technological trigger could include early proof-of-concept proposals, early product manufacturing, and/or no commercial viability, but must trigger public, traditional media, or social media sufficient to shift public interests. Likewise, the immediate "peak of inflated expectations" results when the number of successful or interesting stories begins to rise. However, as the success stories rise, the number of failures or disinterests also rise, which causes the drop into the "trough of disillusionment" that marked the period of waning interest and spontaneous experiments and adaptation from early adopters [4].

After the trough of disillusionment, considered technologies shift back in a positive direction toward the "slope of enlightenment." This upward slope is defined by instances of more beneficial applications of the noted technology with second- and third-generation version of the technology broaching the market and stabilizing the commercial viability. The last noted phase of the so-called hype cycle is a "plateau of productivity," which comes as a slight transition from the previous slope as full mainstream adoption of the considered technology begins to occur. Specifically, the practical and commercial viability becomes clearly defined as brands related to the product establish market relevance [4]. All technologies that emerge do not necessarily push through the spectrum of stages as described within the hype cycle, as some ultimately may fail to become commercial and truly disrupt or revolutionize their application within broader society.

BIOTECHNOLOGY

The first emerging technology that will be evaluated in this chapter is biotechnology. At its foundation, biotechnology is simply technology based on biological systems, components, and resources. More specifically, biotechnology processes harness cellular and molecular processes to develop technologies and products that improve lives and/or ecosystems [5]. For example, modern biotechnology has provided breakthrough products and systems to combat debilitating and rare diseases, address environmental impacts, and create cleaner and more efficient industrial manufacturing processes. For perspective, there are more than 250 biotechnology health care products currently available, 13.3 million farmers using agricultural biotechnology systems, and more than 50 biorefineries for biofuels and chemicals from environmentally friendly renewable biomass [5] (see Figure 4.2).

Biotechnology is divided into five major branches or categories. With the exception of a branch for bioinformatics, the remaining biotechnology branches are referred to as color—most often red, white (sometimes gray), green, and blue [6]. Red biotechnology is simply all biotechnological processes applied to medical processes. Examples include the use of organisms for the production of drugs or antibiotics, use of stem cells for restoration of damaged tissue cells, tests to diagnose certain diseases, and

Figure 4.2 Biotechnology developments are aiding in the development and efficient availability of fuels for cars and other vehicles. (Source: FEMA/Leif Skoogfors.)

ultimately gene therapy. The genetic modification of microbes is far more accepted than the genetically modified food that will be discussed later when reviewing green biotechnology. The continued use and development of red or medical biotechnology has and will continue to serve as a critical element in emergency preparedness efforts for epidemic and pandemic disease outbreaks since biotechnology generation of vaccination is a primary method for successful hazard mitigation [7]. On the other hand, although unlikely, the use of medical biotechnology (particularly the recombinant DNA process) would also potentially be used to create man-made weaponized biological agents, which could be highly dangerous if available to terrorist groups or rogue nations [8].

The second branch of biotechnology relates to industrial processes and is most often given the designation of white or gray. With white biotechnology processes, organizations are modified and/or designed to use for the production of new chemicals, plastic substitutes, vehicle fuels, and clothing fibers [6]. Of these, perhaps the most impactful to disaster reduction and management is the potential positive environmental impacts of white biotechnology's possibilities related to biofuels and biodegradable materials. This type of material may significantly alter the long-term impacts of climate change and global sustainability programming. However, like many aspects of climate change response, the impacts of white biotechnology can only reach these far-reaching implications with appropriate political and economic stimulations from government and private entities that ultimately support additional innovation, entrepreneurship, and market development [9]. The challenge for emergency managers and public safety operators in regard to white biotechnology is how to support these political and economic strategies as it is often difficult not only to make a connection for local and national politicians, but also to stress an importance that exceeds those more immediate issues in a given community. This issue will be further considered in Chapter 12.

The third branch of biotechnology relates to agricultural processes and is often referred to as green biotechnology. It specifically applies to agricultural processes like the production of pest- and disease-resistant crops and plants as well as disease-resistant animals [6]. For example, green biotechnology programs for plant breeding aim to develop superior plant varieties with characteristics like disease resistance, pest resistance, herbicide tolerance incorporation, high nutritive value, easy harvest, good baking qualities, improved nutritional value, and high productivity through improved and selected hereditary breeding [10] (see Figure 4.3). These types of items are commonly referred to as genetically modified

Figure 4.3 Green biotechnology programs create superior plant varieties that are disease resistant and higher in nutritional value. (Source: CDC/Debra Cartagena.)

organisms (GMOs) and have become highly politicized and prohibited in many countries.

One of the most common forms of GMOs is in the form of foods that contain these types of materials. Specifically, the World Health Organization (WHO) defines genetically modified (GM) foods as "foods derived from organisms whose genetic material (DNA) has been modified in a way that does not occur naturally...[but rather] through the introduction of a gene from a different organization" [11]. Most GM foods are available from plants (e.g., corn products), but future food from GM animals is likely to occur within the next decade (see Figure 4.4). From a positive perspective, future GM food production is aimed at altering nutrient values, reducing allergenic potential, or improving food harvesting efficiency with a particular focus on developing nations [11]. In other words, areas of the world that struggle with famine or food development and maintenance would potentially strongly benefit from GM food developments.

Most individuals (and their elected officials) consider that traditionally bred or grown food is 100% safe as its safety has been determined by the many years of its use. However, there is often an equally opposite question about the safety and viability of GM foods. As such, WHO established a series of guidelines that consider the safety of GM foods by investigating direct health effects (e.g., toxicity), tendency to trigger allergic reactions (e.g., allergenicity), nutritional properties, biological stability of the inserted gene, and unintended consequences from the genetic modification [12].

Figure 4.4 Most genetically modified organisms are plant products such as corn. (Source: USDA/Keith Weller.)

Interestingly, no allergenicity has yet been found relative to GM foods currently on the market. However, there is a real and established concern for a mixing of crops between conventional seeds and GM crops, which is commonly called outcrossing [12]. Outcrossing can lead to a negative impact to biodiversity and ultimately changes to the nutritional and biological components of the naturally occurring foods.

In Other Words...Benefit and Risk from Genetically Modified (GM) Foods

Genetically modified (GM) foods (also called genetically engineered or transgenic food) hold great promise that they may provide one of the solutions to help feed growing world populations. But there are also potentially large, and often not well understood, risks from GM technologies—to the environment in general, and to biodiversity and the functioning of ecosystems in particular.

~Harvard University School of Public Health [13]

Although there is still significant debate about the potential risks from GM and certainly how they balance against the potential benefits to greater society, there are a variety of established issues that are often raised. For example, some animal studies have indicated organ damage, gastrointestinal and immune system disorders, accelerated aging, and infertility from the consumption of GM foods [14]. Likewise, some human studies have indicated that GM foods can leave genetic material behind which can transfer into naturally occurring intestinal bacteria and alter the function or purpose of their behavior [14]. The most specific objection typically referenced by professional medical and health associations is against the use of GM bovine growth hormone due to the fact that the milk from treated cows has an increased level of a hormone that has been linked to cancer [14].

As will be established repeatedly in this chapter and throughout the book, the challenges from green biotechnology like genetically modified foods must be constantly balanced against the strong potential benefits of its use. Emergency managers and community leaders throughout the world must constantly strike a balance between inherent risks of a given hazard, service, resource, or technology and its potential benefits to the given area. For example, for the last several years there has been a growing segment of the population objecting to the use of vaccinations (see Figure 4.5). This so-called antivaccine movement (AVM) is objecting to the use of vaccines for preventable diseases like mumps, measles, and pertussis due to concerns about the presence of "toxins" that cause significant side effects (e.g., development of autism) that have been pushed

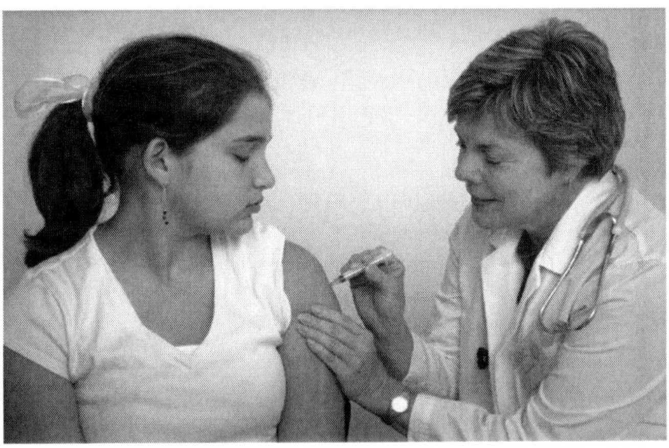

Figure 4.5 There is a growing antivaccine movement which strikes at the balance of risk and reward from various emerging technologies. (Source: CDC/Judy Schmidt.)

in fraudulent research and by outspoken actresses [15]. For example, the state of Wisconsin alone saw more than 7,000 cases of pertussis (aka whooping cough), which was previously nearly eradicated, from 2011 to 2013 [15]. It is this very type of challenge that emergency managers must consider with new and emerging technologies. Is the risk worth the reward? Each technology must be reviewed and considered in this light; however, this challenge is exponentially more difficult when the technology is as global as biotechnology and integrated into altruistic and commercial endeavors.

In Other Words...Risk Versus Reward in Technology

In the United Kingdom, the MMR (mumps/measles/rubella) vaccine is available through the National Health Service. But the MMR vaccine was also the focus of an infamous study by Andrew Wakefield, who suggested that the vaccine had a link to the development of autism.

Although the study was discredited...and struck off the British Medical Registry, his baseless hypothesis pervaded public consciousness, and the country saw a decline in vaccination rates. More recently, a similar antivaccine standpoint has been taken by celebrities...in the United States. [Their] claim, that vaccines contains '"toxins," is less specific than Wakefield's false findings, but also contributed to the misinformation about immunization, leading to a rise in preventable diseases in the United States and other countries such as Australia.

~Rich McCormick [15]

THE INTERNET OF THINGS

Another emerging technology is the integration of the Internet and other data streams into a variety of manufactured and commercial products across all professional disciplines and personal usages. This connectivity to the Internet allows for digital representation and can ultimately be controlled from anywhere with an Internet connection (e.g., mobile devices). Additionally, this connectivity also means more data gathered from more places, which can ultimately improve efficiency and effectiveness of both individual and global safety and security [16]. Given the ubiquitous nature of social media and mobile devices as well as the changing expectations of how communication works (see Chapters 1 and 2), the rise of the Internet of Things is both a natural extension and the next step in the integration of all digital information.

In Other Words...The Rise of the Internet of Things

> As the Internet subsumes physical objects, the rate of change is accelerating...[with] stable wireless platforms, standardized software interface components and cheap, widely available sensors have made the connection of virtually every device—from coffee pots to cars—not only possible; they have made it certain.
>
> ~Jim Stogdill [17]

Because the Internet of Things allows connectivity beyond an individual device, this is a significant departure from the traditional Internet. While specific estimates are wide and varied, the Internet of Things is projected to account for 1.9 billion devices in 2014 and more than 9 billion devices by 2018. By 2018, the number of Internet of Things will roughly equal the number of smartphones, smart TVs, tablet devices, wearable computers, and personal computers combined [18]. Put another way, some experts have projected that it takes approximately 1 trillion sensors to power the billions of connected devices that are expected to power the future of the Internet of Things [19]. While examples of the Internet of Things are wide and variable, the most common and potentially impactful examples include the integration of smart components in kitchen and home appliances, lighting and heating products, safety and security monitoring, health and fitness products, intelligent traffic management systems, waste management systems, and other industrial applications on assembly lines, factories, and warehouses [20] (see Figure 4.6).

The devices and components classified as the Internet of Things typically fall into two categories: sensors and controllers. Sensors are components that monitor any measurable component such as temperature, location, power, flow, radiation, pressure, etc. More complex sensor monitors can also utilize visual and audio triggers in the surrounding environment to initiate controller responses. Furthermore, in some cases these sensors can overlap and connect with one another to further collect and/or analyze environmental data [21]. Conversely, controllers are just as important for devices classified as the Internet of Things. These controllers

Figure 4.6 The integration of the Internet of Things will include various appliances such as these refrigerators. (Source: FEMA.)

are the components that engage with the broader environment around them. They often come in the form of actuators, switches, servos, valves, turbines, and ignition systems. Depending on the system or setup, the sensor, controller, or both can access the Internet through direct linkage or more commonly through wireless access points [21].

In Other Words...The Internet of Things Defined

The Internet of Things refers to the general idea of things, especially everyday objects, which are readable, recognizable, locatable, addressable, and/or controllable via the Internet. These everyday objects include not only electronic devices and the products of higher technological development such as vehicles and high-tech equipment, but things that many people do not ordinarily think of as electronic at all—such as food, clothing, and shelter; materials, parts, and subassemblies; commodities and luxury items; landmarks, boundaries, and monuments; and all the miscellany of commerce and culture.

~FEMA Strategic Foresight Initiative [22]

From an emergency management, homeland security, and public safety perspective there are potentially tremendous benefits from the Internet of Things as well as many potential security pitfalls. For example, some forward-leaning public works officials have begun to consider the integration of sensors into bridges to monitor traffic flow and better understand the pace and severity of common deterioration. This type of functionality would be a great benefit to preventative and mitigation strategies within a community to both identify an issue before it happens and better prioritize the limited funds available to support these efforts [22]. Likewise, search and rescue efforts may ultimately be improved as sensors in phones, clothing, bracelets, and other personal items become more common, which would ultimately serve to "ping" missing individuals in debris and damage after an emergency or disaster [22]. There are numerous other opportunities for emergency preparedness and personal safety, like the so-called "smart" NEST smoke detector discussed in previous chapters (see Figure 4.7).

Unfortunately, with these potential benefits to emergency management and preparedness also come additional risks. Because many of these integrated sensors are applied to everyday items to primarily improve functionality, the robustness of the device security and the information

Figure 4.7 The NEST thermostat and smoke detector are examples of growing "smart" products.

exchanged via the devices are often not as highly developed. Moreover, the criticality of some of these sensors and processors creates additional elements of vulnerability. For example, if critical infrastructure facilities begin to more widely adopt these types of technologies (which all indications show that they are), smart devices and their controllers might be the target of cyber terrorist attacks and ultimately create significant vulnerabilities in critical systems (see Chapter 13). Likewise, government use of Internet of Things devices is already raising concerns about privacy. Individuals are objecting to the possibility that their individual information or data-connected behaviors will be tracked and leveraged in ways they may or may not be aware of [22].

WEARABLE TECHNOLOGY

One of the fastest growing forms of the Internet of Things is wearable technology. Wearable technology is simply the use of a small computer that can be worn by an individual to gather, manipulate, and/or store data. With this broad definition, personal electronic devices as far back as the 1980s that contained added layers of capability (e.g., calculator watches) could be deemed wearable technology [23]. However, clearly the definition of wearable technology—like the devices before it—has changed as the diversity and impact of these devices have broadened in scope and purpose. By today's standards, wearable technology has been applied for commercial usage via uniquely designed interfaces, augmented reality, pattern recognition, and electronic textiles [24]. With 2013 revenue of $8 billion that is expected to more than double by 2017, the industry of wearable technologies is no longer a poorly defined concept of the past, but fully an emerging technology of the future [25].

Perhaps the most well-known (or notorious) form of wearable technology is Google Glass. Like other generalized products such as Kleenex (for facial tissue), Google Glass in many ways has become the only recognizable commercially available form of an augmented reality-based wearable technology. As the basis of the leveraged technology found in Google Glass, augmented reality is a "live, direct or indirect, view of physical, real-world environment[s] whose elements are augmented by computer-generated sensor input[s] such as sound, video, graphics, or GPS data" [26]. At nearly $1,500 per device, Google Glass is not truly available to the masses, but does show tremendous potential in a wide variety of functions. For example, there currently are proposed or tested Google Glass functions in industries including health care, aviation, language translation, navigation, disability access, manufacturing, and law enforcement [27].

In Other Words...The Benefit of Augmented Reality

With the help of advanced augmented reality technology (e.g., adding computer vision and object recognition) the information about the surrounding real world of the user becomes interactive and digitally manipulated. Artificial information about the environment and its objects can be overlaid in the real world.

~Mashable [26]

Emergency management and public safety sectors may benefit from many of these applications. For example, facial recognition capabilities of the augmented reality component might significantly increase the efficiency and effectiveness of security checks and routine safety evaluations at airport security stations and during law enforcement investigations. Likewise, Google's integrated language translation capabilities may also serve to improve the capabilities of humanitarian disaster responders in disaster zones and supplement local translation issues during mass care events in an impacted community. Lastly, there are strong possibilities that the capabilities included in Google Glass could ultimately be leveraged in damage assessment and debris removal activities to track findings and maintain accountability throughout the process. Unfortunately, the Google Glass technology is still new and mostly underdeveloped for commercial and operational applications. There are numerous suggestions and considerations on how this type of wearable technology can be deployed, but the widespread application is still limited. Moreover, given the high cost of the individual devices and the public dislike and distrust of individual users of Google Glass, it will take time for their use to truly become impactful [28].

On the other hand, one form of wearable technology that has successfully been commercially and culturally adopted is the so-called fitness band. Fitness bands are technological gadgets—often worn on the wrist—that collect individual health data, including daily steps (pedometer), heart rate, sleep patterns, and calorie tracking. Research has shown that self-tracking this type of data improves the health of the individual by directly and indirectly encouraging healthier eating, sleeping better, and exercising more [29]. While these devices are also expensive (but far less so than Google Glass), the use by individuals may ultimately be extremely impactful to emerging health threats like rising obesity levels and related conditions like diabetes and heart disease. While these issues will be discussed later in this book, it is important to note that focusing on the impact of individual choices is not new in emergency preparedness, but rather will continue to be a critical element of community preparedness in the future.

DRONES

Unmanned aerial vehicles (UAVs) are an additional technology that has become a prolific and disruptive tool in homeland security, antiterror, and public safety campaigns throughout the world (see Figure 4.8). Drones, as UAVs are more commonly called, are unmanned aircraft that are controlled remotely by a pilot via a ground control system or that have autonomous capabilities based on preprogrammed flight data [30]. Drones are leveraged in situations like reconnaissance or attack missions where manned flight is considered too risky, difficult, or impactful to the mission and goal [30,31]. With the capability to stay airborne for as long as

Figure 4.8 Unmanned aerial vehicles (UAVs) like this MQ-9 Reaper drone are quickly becoming valuable commercial and military tools. (Source: US Air Force/ Staff Sgt. Brian Ferguson.)

17 hours at a time, drones are already highly utilized for militaristic and homeland security functions [30].

Although the United States does not often confirm specific details of drone operations, it has been confirmed that hundreds of terrorist militants have been killed in Pakistan, Afghanistan, and Iraq since 2001 from direct or indirect drone usage. In fact, drones flew more than 100,000 flight hours in support of Operation Enduring Freedom and Operation Iraqi Freedom in 2005 [31]. American use of drones started under President George W. Bush, but has continued to regularly occur under the supervision of President Barack Obama (see Figure 4.9). The precision and effectiveness of drone intervention have never been questioned; however, the legality, constitutionality, and appropriateness of use of drones for targeted intervention have been questioned as the decision to use drones often circumvents traditional war clauses within democratic communities [30].

As is often the case in technology first used for militaristic purposes, there is a growing movement toward utilizing drones for public safety, homeland security, and emergency management activities. For example, after the massive typhoon Haiyan struck the Philippines in 2013,

Figure 4.9 American use of military drones started under President George W. Bush and continued under current President Barack Obama. (Source: White House/Eric Draper.)

drones were used to help identify areas in need of debris removal and to help locate bodies in debris [32]. Likewise, drones have been used after disaster events in Peru, Japan, and Haiti to produce aerial imagery for improvement of situational awareness after the event [33]. Similarly, an Australian fire department recently announced that drones would be used to monitor wildfires and other major events. Specifically, the organization is using remote-controlled drones to take aerial images that are transmitted back to incident controllers in real time [34]. Clearly, forward-leaning emergency preparedness and response organizations are just beginning to scratch the surface of the potential of drones in emergency response and recovery efforts.

Unfortunately, like all the emerging and disruptive technologies discussed in this chapter, the use of drones in emergency response and homeland security issues has limitations and challenges. First, there are technical limits to how well drones can maneuver in tightly packed urban and suburban environments that often lack any consistency of physical or structural patterns that could be leveraged by the technology [35]. Second, the aviation management sector has yet to fully grasp the long-term management of the personal, commercial, and governmental uses of drones. Issues related to training, operational specifications, and technological issues plague oversight groups like the US Federal Aviation Administration (FAA), which thus far has been reluctant to update management protocols at the same speed at which the drone technology develops and is implemented [36]. The final major challenge for drone implementation is personal privacy. Without a doubt, people are concerned about the possible misuse or application of drones, particularly in domestic applications such as law enforcement. This debate has driven as many as 11 states to consider restrictions on the use of drones "over their skies amid concerns that the vehicles could be exploited to spy on Americans" [37]. Other larger municipalities have suggested allowing drones in law enforcement, but not over large assemblies of people outdoors or for nonspecified surveillance activities [37].

In Other Words...Drone Management and Related Safety Issues

Because [drones] are inherently different from manned aircraft, introducing [drones] into the nation's airspace is challenging for both the FAA and aviation community. [Drones] must be integrated into a National Airspace System (NAS) that is evolving from ground-based navigation aids to a GPS-based system in NextGen. Safe integration of [drones] involves gaining a better understanding of operational issues, such as training requirements, operational specifications and technological considerations.

~US Federal Aviation Administration (FAA) [36]

THREE-DIMENSIONAL PRINTING

Three-dimensional (3D) printing is one of the most interesting emerging technologies that are already having a distinctly disruptive impact on a variety of sectors including public safety and emergency management. According to one source, 3D printing is the building of "objects from a three-dimensional digital file by laying down an ultrathin layer of whatever material or materials the object is made of, and then add[ing] each additional ultrathin layer—one by one—until the object is formed in three-dimensional space" [38]. Three-dimensional printing is also sometimes referred to as additive manufacturing as it adds layers of material to build an improved product, which is in contrast to traditional manufacturing techniques (subtractive processes) that mostly rely on the removal of material (e.g., cutting) to achieve an improved state [39]. However, given the relatively low cost of materials and printers, numerous industries, including jewelry, footwear, architecture, engineering, automotive, medical, and many others, are already leveraging the capabilities of 3D printing.

The potential use for this type of printing also has significant potential in disaster recovery and public safety sectors. For example, for approximately $5,000, a Chinese company is harnessing 3D printing technology to build one-story houses that are 10 meters wide and 7 meters high, using a mix of cement and construction waste. Likewise, researchers at the University of Southern California developed a 3D printer capable of building a 2,500 square-foot house within 24 hours [40]. If this technology continues to develop and remain affordable, this type of application may quickly become an easy solution to address lack of substantial housing in developing countries and as temporary housing in disaster recovery areas (see Figure 4.10). For example, the Disaster Housing Assistance Program (DHAP) in cooperation with the Federal Emergency Management Agency (FEMA) and the Department of Housing and Urban Development (HUD) spent more than $820 million to provide housing to

Figure 4.10 As 3D printing becomes more developed and practical, it may allow for efficient and affordable production of housing for disaster recovery. (Source: FEMA/Elissa Jun.)

approximately 60,000 households as part of the recovery from Hurricanes Katrina and Ike [41]. With an average cost of more than $13,000 per household, traditional housing manufacturing is far more costly than currently produced 3D printed homes.

On the opposite end of the spectrum, there is growing concern about the impacts to public safety that may rise due to the seemingly limitless printing capabilities that exist for this particular technology. For example, a Texas firm has produced a metal pistol based on a Browning 1911 model firearm via a 3D printing process. Specifically, the company utilized a printing process already leveraged in the application of aerospace and medical industries so it was well suited to the precision needed to create the firearm [42]. This particular gun fired as many as 50 rounds to show viability and long-term reliability [43]. Unlike previously printed plastic weapons that were illegal under the Undetectable Firearms Act of 1988, the metal components not only are legal, but can also be produced by the printer as a legitimate weapons manufacturer [43]. However, other print designs for nonmetal guns have been widely available on the Internet with more than 100,000 downloads. Likewise, US Alcohol, Tobacco, and Firearm (ATF) agents who tested the nonmetal design found that the weapon had the power to cause lethal damage in humans [44]. Given the overall flexibility of these printing models, it is unclear how government regulators will handle the licensure and accountability of 3D printed weapons into the future.

AUTONOMOUS CARS

The idea of self-driving or autonomous cars seems like science fiction and beyond even the scope of a book focused on the future. However, the reality of a fully self-driving car is not as far off into the future as it sounds. According to a 2014 report, one or more commercially available vehicles can already autonomously provide lane departure warnings, blind spot monitoring, forward collision warnings, adaptive cruise control, and pedestrian detection [45]. However, these features are merely scratching the surface of what may be possible and how emergency response and particularly public safety may be improved with the acceptance and integration of this type of technology.

In Other Words...The Challenge of Accepting Self-Driving Cars

Getting a car that can operate with the reliability that today's cars do is a staggering challenge...[but] even if autonomous cars save countless lives, one headline [reading] "Machine Kills Child" trumps [saved lives].

~Bryant Walker Smith, Center for Automotive Research at Stanford University [45]

The most well-known example of this technology is Google's (sort of) secret Google X Lab driverless car project, which the company has stated will be publicly available by 2017. Currently, Google's driverless car can navigate up to 1,000 different urban environment conditions such as bicyclists, construction zones, and tight vehicular traffic. Google's concept is to use smart sensors (like those discussed earlier in this chapter), including radars and lasers, to create a three-dimensional map of a self-driving car's surroundings in real time while the autonomous software categorizes the data into four categories: moving vehicles, pedestrians, cyclists, and static items (e.g., parked cars and road signs). According to Google engineers, their vehicles have driven without a driver for more than 700,000 miles since testing began on both interstate and surface streets [46].

Automotive industry experts are projecting annual sales of driverless vehicles to balloon to 11.8 million by 2035. Over the course of time that would project out to more than 54 million cars on the road that needed minimal or no driver input. Unfortunately, the estimated cost of these vehicles will initially cost an additional $7,000 to $10,000 per car as early as 2025, but should steadily drop after the initial spike [47]. While these projects are still 20 or more years out, it is clear that self-driving cars are coming along, will have both personal and commercial applications, and may ultimately reduce injuries and fatalities on the roads.

This type of technology is not being developed to simply create another line of products for automotive companies, but rather has a primary purpose to improve safety. For the last 20 years, there has been an average of more than 10 million automobile accidents in the United States each year [48]. Motor vehicle crashes are the leading causes of death in the United States, with more than 2.3 million adult drivers and passengers per year being medically treated in hospitals due to being in an accident. Likewise, the lifetime costs of crash-related deaths and injuries among drivers and passengers are well above $70 billion [49]. While the causes of automobile accidents are varied, the impact of human choices and bad decisions has long been a primary cause. This is even truer than ever as people continue to bring technology and the fast pace of life into the automobile with them (e.g., texting and driving). Consequently, the integration of autonomous safety features like facial recognition cameras to avoid sleepy behavior or the nearly complete removal of human choice in the process may vastly improve the overall safety of mass transportation in the future and ultimately improve transportation efficiency.

VIRTUAL REALITY

The last emerging technology that will be considered in this chapter is virtual reality. Virtual reality is a term used to describe the technology that utilizes a three-dimensional, computer-generated environment which can be explored and interacted with via head-mounted displays [50]. This interaction creates a simulated world where the user can engage and manipulate objects in the virtual environment. Some virtual reality systems add additional sensory components such as sound or overlaid multimedia [50]. While virtual reality has received significant additional publicity these last few years, the concept of virtual reality was first introduced in the 1950s and has been intermediately developed for war training, gaming, and other educational applications since that time [51].

In Other Words...Describing Virtual Reality

Virtual reality is the ultimate medium of syntactical intent; the only way to figuratively "show" someone exactly what you mean is to literally show them. Words are exceptionally ineffective at conveying meaning, as they are a low-bandwidth...medium of knowledge transference. Virtual reality will let us remove the ambiguity that is the discrepancy between our internal dictionaries and bypass communication through symbolism altogether. The result will be perfect understanding, as all parties behold the same information.

~The Future of Virtual Reality [52]

Even though the virtual reality technology has been in development for much of the last 50 years, it was often fraught with issues that would limit widespread and continued usage. For example, image latency, blurry images, and often buggy technological consistency created frustration or physical challenges including nausea [53]. These challenges have traditionally been created through limitations in the human brain to accept the information displayed in the virtual environment and limitations in the display and device optics. For example, virtual reality devices must constantly and without interruption balance the information from a built-in gyroscope, accelerometer, and magnetometer to report locations and perspective within this world [53]. Unfortunately, major technology observers and media publications have routinely predicted various virtual reality technologies as the breakthrough to handle and address these complicated functions, but have routinely been shown to be false [53].

With all those limitations aside, an American company called Oculus VR was founded in 2012 and has yet again raised interest in and awareness of the possibilities of the virtual reality technology. After initially raising $2.4 million via crowdfunding on Kickstarter to develop an Oculus Rift headset for gaming, social networking giant Facebook purchased the company—and related technology—for $2 billion [54]. Upon announcing the deal, Facebook founder Mark Zuckerburg stated that the intention was to move beyond simple gaming applications with the ability to enjoy "a courtside seat at a game, studying in a classroom of students and teachers all over the world or consulting a doctor face-to-face—just by putting on goggles in your home" [54].

While Oculus Rift is not the only virtual reality technology under significant development, the overall potential for this technology within the emergency management community is potentially profound—although still on the horizon of the future. As another communication process, this type of technology presents a significantly interactive possibility for the delivery, acceptance, and behavioral change patterns that are challenging before, during, and after emergencies. Virtual reality technologies may also help bring together geographically diverse or disaster-isolated individuals into virtual discussions, discourse, or town hall-type environments, which are again difficult to efficiently produce in real-world disaster environments [53]. This same connectivity could also improve ongoing and just-in-time training necessary before responding or coordinating recovery from an event. The last and perhaps most forward-looking application of virtual reality is the use of these technologies to supplement rest and relief efforts for disaster responders. Taking the proper downtime to rest and recover is often extremely difficult for emergency responders for a variety of personal, sociological, and psychological reasons, but has long been established as a best practice in disasters. If virtual reality could provide the sensation and escape necessary to recovery while never leaving the disaster zone, the recovery and long-term sustainability of these responders might be vastly improved (see Figure 4.11).

PUBLIC PERCEPTIONS OF TECHNOLOGY

Regardless of the current and potential benefits of the emerging and disruptive technology, the public perception of said technologies will play a role in the pace, effectiveness, and, ultimately, longevity of these systems. According to a recent Pew Internet study, most Americans anticipate that the technological developments of the next half-century will have a net positive impact on society. Specifically, 59% of those surveyed were optimistic that emerging technological changes will make life better, while 30% stated that they felt that life would be worse than today. Interestingly, of those emerging technologies discussed in this chapter, 50% of those surveyed said they would not be interested in riding in a driverless car and only 20% indicated they would be interested in lab-grown foods.

Figure 4.11 Virtual reality systems can often be valuable tools for responders and military personnel to receive the effects of events before they happen. (Source: US Navy.)

These perceptions of technology did not significantly differ based on age, but the overall positive optimism did increase as income increased [55]. This public perception of technologies may ultimately hinder or help the progress of these emerging technologies as sources of aid to the emergency management community.

REFERENCES

1. Santoso, Alex. (2012). "Four Geeky Laws That Rule Our World." Neatorama. http://www.neatorama.com/2012/09/05/Four-Geeky-Laws-That-Rule-Our-World/#!K7lg2. Accessed May 8, 2014.
2. "The Road to Self-Driving Cars." (2014). *Consumer Reports*. April 2014. pp. 16–20.
3. Sparshot, Jeffrey. (2013). "About 1 in 3 Households Has No Landline Phone." *Wall Street Journal*. http://blogs.wsj.com/economics/2013/09/05/about-1-in-3-households-has-no-landline-phone. Accessed May 8, 2014.
4. "Hype Cycles: Interpreting Technology Hype." (n.d.) Gartner. http://www.gartner.com/technology/research/methodologies/hype-cycle.jsp. Accessed May 9, 2014.
5. "What Is Biotechnology?" (2014). Biotechnology Industry Organization. http://www.bio.org/articles/what-biotechnology. Accessed May 9, 2014.

6. "Branches of Biotechnology." (2010). The Agricos. http://theagricos.com/biotechnology/branches-of-biotechnology. Accessed May 10, 2014.
7. Liew, F. W. (1990). "Biotechnology of Vaccine Development." The University of Nottingham. http://www.nottingham.ac.uk/ncmh/BGER/pdf/Volume%208/BGER8-2.pdf. Accessed May 10, 2014.
8. Casagrande, Rocco. (2013). "Technologies That Will Change the World This Century." IAEM Annual Conference Presentation. https://s3.amazonaws.com/v3-app_crowdc/assets/events/11sg2pbbrj/activities/Tech_That_Will_Change_the_World_Rocco.original.1382635841.pdf. Accessed May 10, 2014.
9. "Industrial or White Biotechnology—A Driver of Sustainable Growth in Europe." EuropaBio. http://www.europabio.org/industrial-or-white-biotechnology-driver-sustainable-growth-europe. Accessed May 10, 2014.
10. "Agriculture Biotechnology." (2010). The Agricos. http://theagricos.com/biotechnology/agriculture-biotechnology. Accessed May 10, 2014.
11. "Food, Genetically Modified." (2014). World Health Organization (WHO). http://www.who.int/topics/food_genetically_modified/en. Accessed May 10, 2014.
12. "20 Questions on Genetically Modified Foods." (2014). World Health Organization (WHO). http://www.who.int/foodsafety/publications/biotech/20questions/en. Accessed May 14, 2014.
13. "Genetically Modified Foods." (2012). Harvard University School of Public Health. http://chge.med.harvard.edu/topic/genetically-modified-foods. Accessed May 14, 2014.
14. "10 Reasons to Avoid GMOs." (2013). Institute for Responsible Technology. http://www.responsibletechnology.org/10-Reasons-to-Avoid-GMOs. Accessed May 14, 2014.
15. McCormick, Rich. (2014). "Map of Preventable Disease Outbreaks Shows the Influence of Anti-Vaccination Movement." *The Verge*. http://www.theverge.com/2014/1/21/5329478/vaccine-preventable-disease-outbreaks-show-anti-vaccine-movement-influence. Accessed May 18, 2014.
16. "Internet of Things." (n.d.). Cisco. http://www.cisco.com/web/solutions/trends/iot/overview.html. Accessed May 19, 2014.
17. Martin, Glen. (2014). "How the Internet of Things Is More Like the Industrial Revolution Than the Digital Revolution." *Forbes*. http://www.forbes.com/sites/oreillymedia/2014/02/10/more-1876-than-1995. Accessed May 19, 2014.
18. Adler, Emily. (2013). "Here Is Why the Internet of Things Will Be Huge and Drive Tremendous Value for People and Businesses." *Business Insider*. http://www.businessinsider.com/growth-in-the-internet-of-things-2013-10. Accessed May 19, 2014.
19. Sharma, Rakesh. (2014). "A New Perspective on the Internet of Things." *Forbes*. http://www.forbes.com/sites/rakeshsharma/2014/02/18/a-new-perspective-on-the-internet-of-things. Accessed May 21, 2014.

20. Adler, Emily. (2014). "These Are the Killer Apps That Will Make the Internet of Things Indispensable in Everyday Life." *Business Insider*. http://www.businessinsider.com/these-are-the-killer-devices-and-apps-that-will-make-the-internet-of-things-indispensible-2014-1. Accessed May 21, 2014.
21. Proffitt, Brian. (2013). "The Internet of Things Might Try to Kill You." read*w*rite. http://readwrite.com/2013/09/18/internet-of-things-security-disaster-terrorism-war#awesm=~oELqYc4sQZfN53. Access May 21, 2014.
22. "Strategic Foresight Initiative." (2011). FEMA. http://www.fema.gov/pdf/about/programs/oppa/technology_dev_%20paper.pdf. Accessed May 21, 2014.
23. Racoma, J. Angelo. (2013). "Wearable Tech: A Brief History and Look to the Future." androidauthority. http://www.androidauthority.com/wearable-computing-history-238324. Accessed May 22, 2014.
24. "Wearable Computer." (2014). Technopedia. http://www.techopedia.com/definition/16339/wearable-computer. Accessed May 22, 2014.
25. Sabhlok, Raj. (2013). "Smartwatches, Google Glass, and the Wearable Technology Chocolate Box." *Forbes*. http://www.forbes.com/sites/rajsabhlok/2013/12/05/smartwatches-google-glass-and-the-wearable-technology-chocolate-box. Accessed May 22, 2014.
26. "Augmented Reality." (2014). *Mashable*. http://mashable.com/category/augmented-reality/. Accessed May 22, 2014.
27. Suleman, Khidr. (2014). "Google Glass: Ten Case Uses for Wearable Technology." ITPro. http://www.itpro.co.uk/mobile/21581/google-glass-10-use-cases-for-wearable-technology. Accessed May 23, 2014.
28. Honan, Matt. (2013). "I, Glasshole: My Year with Google Glass." *Wired*. http://www.wired.com/2013/12/glasshole. Accessed May 23, 2014.
29. "Why Use a Fitness Tracker?" BASIS. http://www.mybasis.com/basis-fitness-tracker. Accessed May 23, 2014.
30. "UAV Models." (n.d.). The UAV. http://www.theuav.com. Accessed May 23, 2014.
31. "Drones: What Are They and How Do They Work?" BBC News. http://www.bbc.com/news/world-south-asia-10713898. Accessed May 23, 2014.
32. "Drone Aids Typhoon Clean-Up in the Philippines." (2013). Thompson-Rueters Foundation. http://www.trust.org/item/20131128163513-3o3d6. Accessed May 25, 2014.
33. Meier, Patrick. (2014). "Humanitarians Using UAVs for Post Disaster Recovery." iRevolution. http://irevolution.net/2014/03/26/humanitarians-using-uavs-for-post-disaster-recovery. Accessed May 25, 2014.
34. Webb, Carolyn. (2013). "Drones to Fight Fires." The Age—Digital Life. http://www.theage.com.au/digital-life/digital-life-news/drones-to-fight-fires-20130903-2t3bd.html. Accessed May 25, 2014.
35. Ferenstein, Gregory. (2013). "Three Problems Stopping Bezos' Army of Amazon Delivery Drones." Tech Crunch. http://techcrunch.com/2013/12/02/amazon-delivery-drones. Accessed May 25, 2014.

36. "Fact Sheets—Unmanned Aircraft Systems (UAS)." (n.d.) Federal Aviation Administration. http://www.faa.gov/news/fact_sheets/news_story.cfm?newsId=14153. Accessed May 25, 2014.
37. "Seattle Mayor Ends Police Drone Efforts." (2013). *USA Today*. http://www.usatoday.com/story/news/nation/2013/02/07/seattle-police-drone-efforts/1900785. Accessed May 25, 2014.
38. Gore, Al. (2013). *The Future: Six Drivers of Global Change*. New York: Random House, p. 31.
39. "3D Printing." (n.d.). *Mashable*. http://mashable.com/category/3d-printing. Accessed May 26, 2014.
40. Goldin, Melissa. (2014). "Chinese Company Builds Houses Quickly with 3D Printing." *Mashable*. http://mashable.com/2014/04/28/3d-printing-houses-china. Accessed May 26, 2014.
41. "Effectiveness and Costs of FEMA's Disaster Housing Assistance Program." (2011). FEMA. http://www.oig.dhs.gov/assets/Mgmt/OIG_11-102_Aug11.pdf. Accessed May 26, 2014.
42. Krawczyk, Konrad. (2013). "Texas Firm Makes World's First 3D Printed Gun." *Fox News*. http://www.foxnews.com/tech/2013/11/08/texas-firm-makes-worlds-first-3d-printed-metal-gun/?intcmp=features. Accessed May 26, 2014.
43. Daileda, Colin. (2013). "First 3D Printed Metal Gun Fires 50 Rounds and Counting." *Mashable*. http://mashable.com/2013/11/11/3d-printed-metal-gun. Accessed May 26, 2014.
44. Greig, Alex. (2013). "World's First Fully 3D Printed Gun Is Tested by ATF Agents." *Daily Mail*. http://www.dailymail.co.uk/news/article-2507654/The-worlds-fully-3-D-printed-gun-tested-Feds-blows-faces.html. Accessed May 26, 2014.
45. "The Road to Self-Driving Cars." (2014). *Consumer Reports*. April: 16–20.
46. Pritchard, Justin. (2014). "Google: Our Driverless Cars Are Mastering City Streets." *Business Insider*. http://www.businessinsider.com/google-our-driverless-cars-are-mastering-city-streets-2014-4. Accessed May 27, 2014.
47. Woodyard, Chris. (2014). "Study: Self-Driving Car Sales Will Explode." *USA Today*. http://www.usatoday.com/story/money/cars/2014/01/02/self-driving-study/4292893. Accessed May 27, 2014.
48. "Table 1103: Motor Vehicle Accidents—Number and Deaths." US Census Bureau. https://www.census.gov/compendia/statab/2012/tables 12s1103.pdf. Accessed May 27, 2014.
49. "Motor Vehicle Safety." (2012). US Centers for Disease Control and Prevention (CDC). http://www.cdc.gov/Motorvehiclesafety. Accessed May 27, 2014.
50. "What Is Virtual Reality?" (2009). *Virtual Reality*. http://www.vrs.org.uk/virtual-reality/what-is-virtual-reality.html. Accessed May 28, 2014.
51. "Virtual Reality History." (1995). University of Illinois. http://archive.ncsa.illinois.edu/Cyberia/VETopLevels/VR.History.html. Accessed May 28, 2014.
52. "The Future of Virtual Reality." (n.d.). The Nano Age. http://www.thenanoage.com/virtual-reality.htm. Accessed May 28, 2014.
53. Winters, Dan. (2014). "Inside the Oculus Rift." *Wired*. June: 77–95.

54. Dredge, Stuart. (2014). "Oculus Rift—10 Reasons Why All Eyes Are Back on Virtual Reality." *The Guardian*. http://www.theguardian.com/technology/2014/mar/31/oculus-rift-facebook-virtual-reality. Accessed May 29, 2014.
55. Smith, Aaron. (2014). "U.S. Views of Technology and the Future." Pew Internet. http://www.pewinternet.org/2014/04/17/us-views-of-technology-and-the-future. Accessed May 29, 2014.

Section II

Preparedness, Response, and Recovery

5

Age, Gender, and Sexuality Roles

Disaster risk reduction that delivers gender equality is a cost-effective win-win operation for reducing vulnerability and sustaining the livelihoods of whole communities.

~Margareta Wahlstrom
UN Assistant Secretary-General for Disaster Risk Reduction [1]

MEN AND WOMEN

Looking at trends in technology is only a small fraction of those issues that may ultimately impact emergency management in the future. For example, this chapter will focus on how age, gender, and sexuality roles impact the capability and resiliency of a community to respond to and recover from emergencies and disasters. This consideration will evaluate not only individual decisions related to disaster readiness, but the cultural, societal, ethical, and governmental standards that affect the resources and activities within a given community. These current standards will also be evaluated against current societal trends shifting these standards toward new and undefined areas that may ultimately positively or negatively impact both individuals and their impacted communities to fully recover after a disaster.

Community is defined by two major components: personal values and societal norms. Personal values or perspectives such as individualism or egalitarianism are programmed by early socialization. For example, a person's formative experiences are broadly defined during childrearing, schooling, employment, or other personal experiences (see Figure 5.1). These personal values "operate cognitively like the self-concept, as chronically accessible,

Figure 5.1 A person's formative experiences are strongly defined during childrearing, schooling, employment, and other personal experiences. (Source: FEMA.)

unwavering lenses that shape the person's choices" [2]. Consequently, this concept of personal value does limit the justification of personal beliefs to internalized value orientation rather than any influence from groups or community. In other words, personal beliefs do not necessarily have to be justified or in compliance with broader cultural standards. This limit is where social norms become relevant. Whereas personal values are related to an individual view, social norms relate to beliefs about other people. Within social science studies, social norms more broadly describe what an individual believes and expects from others within the community or society. Moreover, social norms are often divided into descriptive norms, which describe the prevailing community attitudes, and injunctive norms that refer to what ought to be accepted. Interestingly, unlike personal values, cultural norms are often far more malleable to shifting societal changes [2]. For example, public opinions about racial and gender equality have often changed at a slower pace than broader social norms.

The impact of social norms has long been highly impactful on disaster resilience as the broader community provides emotional support and physical infrastructure to support recovery. Consequently, it will be the primary

Figure 5.2 Cultural norms have changed in many parts of the world regarding how women are perceived in comparison to male counterparts. (Source: FEMA/Brittany Trotter.)

basis of how age, gender, and sexuality roles are shifting and will ultimately need to be measured by the emergency management community. These shifting cultural norms are ultimately not universal as the collection of society in various parts of the world includes ranges of socioeconomic, religious, educational, and geographic characteristics that will create a range of social norms and a different perspective. For example, in many countries like the United States and much of Europe, the cultural norm for the role of women within society has shifted away from solely a maternal homemaker to include significant strides in working and economic equality with men (see Figure 5.2). This contrasts significantly with many developing countries, where the same societal norm remains built around more traditional domestic roles that often coincide with limited education and experiences.

In Other Words...Impact of Disaster Vulnerability on Gender-Based Inequalities

Two elements in combination increase or decrease disaster risk: a potentially damaging event or phenomenon (hazard), and the degree of susceptibility of the elements exposed (vulnerability). A natural hazard like a hurricane or earthquake will pose a greater or lesser risk, depending on its own physical intensity and the vulnerability and capacities of the people exposed to the hazard. The negative impact of risk therefore depends on the characteristics and intensity of the hazard. Gender-based differences and inequalities have a strong negative or positive effect on the vulnerability and capacities of people exposed to hazards. Gender relations will determine the magnitude of the risk.

~United Nations Report on Disaster Risk Reduction [3]

Disaster sociologists have long noted the disproportionate impact of disasters on women—particularly in developing countries. For example, one United Nations study noted that 90% of the 140,000 fatalities in Bangladesh from Cyclone Gorky in 1991 were women [3]. An additional survey showed that as many as 71% of women in flood-impacted areas experience physical harm or punishment during disaster events [4]. Moreover, these findings are not limited to the impact of disasters on developing countries. Specifically, a 2007 study of Hurricane Katrina found that the vast majority of those living alone were women who were less likely to have cell phones or access to computers, which significantly added to the community impact from that particular event [5] (see Figure 5.3). A 2003 heat wave in Europe revealed more women dying—particularly elderly women—than men [6]. These studies are strong indicators that the cultural roles and expectations of women in communities throughout the world have increased the risk with clear indications that the capability to recover from an emergency or disaster of any size or scope may be hindered.

While each geographic area has specific considerations of women within their community, there are some commonly agreed-upon challenges that impact the risk to women globally. These preparedness challenges include physical aspects, social and cultural aspects, limited education, poorer health, and economic engagement. For example, a 2005 study in Sri Lanka after a tsunami determined that women were at greater risk because knowing how to swim and climb trees was often only taught to boys [6]. While developed and more progressive countries would scoff at some of these limitations in light of gender equality movements, some are inherently impactful regardless of political or social progression. For example, the noted physical limitations also became evident when considering the ability to mitigate physical risk in homes and/or

Figure 5.3 A study found that Hurricane Katrina impacted women disproportionately due to statistically less access to cell phones and computers. (Source: FEMA/Patsy Lynch.)

residences (e.g., applying hurricane shutters). Likewise, the traditional cultural concepts related to maternal roles often limit educational opportunities or continued commitment to these physical mitigation processes.

The commonly identified process to address these additional risk factors is the concept of gender mainstreaming. This concept is a global approach to promote gender equality before, during, and after disasters as well as for other social and cultural circumstances. Since 1997, the United Nations assistant secretary-general and special advisor to the secretary-general on gender issues and the advancement of women has been tasked with supporting gender mainstreaming through activities including policy development, research, advocacy, legislation, resource allocation, and planning [7]. Given this approach, there have been some measures of success from gender mainstreaming including international gender policies and action plans, gender equity in social justice missions, some poverty alleviation, greater participation of women in community projects, and some increase in political will toward the improving equity of women in relationship to men [8].

However, like any major international push, gender mainstreaming has not been easily integrated into many global communities. One report devalued the minimal successes as only creating "islands of success" [9]. These limitations have been associated with various reasons, but generally are categorized into three main areas. The first major limitation to the effectiveness of gender mainstreaming is a lack of information regarding the benefits of gender integration being available to all communities throughout the world. Likewise, a second limitation is a lack of awareness of gender issues in leadership within all global communities. The third limitation is a lack of women in community leadership positions. While these limitations are primarily defined for all gender-mainstreaming applications, they also have a significant impact in risk reduction throughout the world. For example, the World Bank conducted a case study in 2009 of the impact of Typhoon Ondoy and Tropical Storm Pepeng on the Philippines (see Figure 5.4). This study found that gender mainstreaming efforts had not been effective as gender inequality was still highly impactful during the event. In response, the Philippine government created a gender-sensitive framework strategy related to climate change where it specifically acknowledged women as a vulnerable group [9].

Figure 5.4 Gender mainstreaming was not effective during Typhoon Ondoy and similar events and thus had disproportionate effects on recovery and restoration in the community. (Source: US Navy/Lance Corporal Marie Matarlo.)

The global commitment to gender mainstreaming and gender equality still has many hurdles and unfortunately is not broad in scope. Consequently, it often falls outside the immediate concern of most emergency management and risk reduction personnel. These issues are also vastly different from one part of the world to another as one community may still firmly hold to traditional cultural or even tribal roles for women, while more developed countries may ostensibly allow for equal opportunity for men and women. Regardless, the continued commitment to community or cultural risk reduction through minimizing the additional vulnerabilities created through gender inequalities must continue to be a focus not only with international groups like the United Nations, but also with state, regional, and even local emergency management professionals who will keenly understand the circumstances and roles of women and related risk in their particular communities.

In Other Words...Shifting Gender Perspectives in Risk Reduction Strategies

There has recently been a critical shift in the mainstreaming of gender perspectives in disaster risk reduction; from a women-focused approach to a gender-focused approach, based on the premise that the roles and relationships of women and men in disaster risk reduction should be analyzed within the overall gendered socioeconomic and cultural context. On top of this shift, the strategic focus of disaster management has changed from reactive disaster response to long-term proactive disaster risk and vulnerability reduction where gender and disaster risk reduction are considered necessary.

~The World Bank [10]

Clearly, some of these cultural and community norms are shifting—particularly in industrialized nations like the United States. For example, according to a 2013 *USA Today* survey, 28% of wives outearn their husbands when both work, which is nearly twice as much as 25 years prior. To put it another way, more than 12 million American families have a female "breadwinner." This shift has been exemplified by Facebook Chief Operating Officer Sheryl Sandburg and Yahoo Chief Executive Officer Marissa Mayer, who have written and spoken on these issues as well as implemented internal organizational policies related to the complex choices women may face based on current cultural norms. The increased presence of women in high-paying and high-profile positions as well as throughout the American workforce reinforces the importance of education, economic stability, and health protection. This shift is no more evident than in the gender breakdown of college degree recipients. Specifically, according to 2010 US Department of Education data, women earned 57% of bachelor's degrees, 60% of masters degrees, and 52% of doctoral degrees [11]. This current trend and future shift are critical to broad risk reduction given that disaster research in 141 countries revealed that there was a direct correlation in the number of disaster-related deaths for women and a woman's lack of economic and social rights within a given society. Specifically, the same study found that in societies where women and men both enjoyed equal (or near equal) rights, disasters caused the same number of deaths in each gender [6].

While the role of women in disasters and related risk reduction has been heavily studied and needs targeted intervention to succeed, the role of men in disasters should not be ignored. Not only does their engagement with women during disasters play a role in increasing or decreasing risk in a given event, but there are also unique characteristics of male behavior that must be noted and observed. For example, a 1994 study showed that males (in this particular case, Caucasian males) consistently exhibited lower perceptions of risk across a wide variety of societal hazards [12]. This was reinforced by another study between 1994 and 2000 that found American men were more than twice as likely to die in a thunderstorm or be struck by lightning [13]. Likewise, disaster researchers have found that increased obesity rates and drug usage among Russian males have increased their susceptibility to extreme heat conditions [14]. Most researchers agree that these increased risks relate to the willingness of men to work in more hazardous conditions and a willingness to underestimate the actual level of risk in a given situation (see Figure 5.5).

In Other Words...Gender-Based Perception of Risk

As men and women view the world differently, it follows that they will also perceive risks differently. Women are ambivalent about taking risks, while men view risks and hazards as part of life. Indeed, since women are more concerned about destructive technologies and war and men more prone to aggressive or risk behaviors, it may be generalized that men are "risk-takers" and women are "risk-avoiders."

~Alice Fothergill [15]

These tendencies for men to have increased risk in certain hazardous situations are not limited simply to uncontrollable circumstances. For example, one research study showed that while there was no distinct difference between the disaster preparedness efforts of men and women, men showed an increase in interest in the technical aspects of preparedness or prevention before a disaster [15]. This characteristic is an interesting development given the rise of technology and the use of mobile and smart devices. Access to technical information from preparedness sources, including news media, emergency management organizations, and a variety of other digital sources, is widely available and potentially could

Figure 5.5 Research has consistently shown men to have lowered perceptions of risk due to increased willingness to be exposed to hazardous conditions. (Source: FEMA/Jocelyn Augustino.)

impact how men accept and act on disaster-related information. Similarly, a separate study revealed that before Hurricane Andrew in 1992, women were responsible for preparing their family members (e.g., stocking supplies) while the men were responsible for the external preparedness of the residence [15]. Interestingly, as traditional domestic roles have changed (as described earlier) and women's roles in industrial nations have expanded, it is unclear whether this preparedness pattern would still exist given the age of the findings.

Traditionally, men and women have also perceived risk communication in vastly different ways. Specifically, women are more likely to hear warnings from peers such as friends, neighbors, and relatives and act on this socially confirmed warning of risk. Conversely, men are often skeptical of socially confirmed warnings and are therefore less likely to take disaster prevention information seriously [15]. This role of socially confirmed risk communication is again particularly relevant in the age of social media and instantaneous communications. According to one study, women are more likely to use social media for relationships, information sharing, and self-help than men. For example, 65% of those women surveyed utilized social media to stay in touch with family and friends, while only 53% of men used social media for such purposes [16].

The most significant hindrance to the preparedness of men for disasters is traditional views of masculinity. In Western culture, this type of masculinity is often characterized by heterosexuality, aggression, authority, courage, decision making, rationality, emotional control, muscular prowess, risk taking, and dominance. These concepts of so-called "manliness" in industrial communities and traditional cultural roles for men often mean men are less likely to access health services because they are concerned that the broader community might perceive them as weak and therefore hinder their willingness to receive health support or government aid in a disaster [17]. This type of behavior is also one of the primary reasons men take more risks than women as discussed earlier. The drive to maintain these preconceived personal standards and cultural norms can lead to poor choices and increased risk.

It is important to note that these traditional gender roles may be shifting. For example, one study after the 2008 American recession found that male-dominated industries like construction were harder hit than female-dominated industries. Consequently, male unemployment was almost 2% higher than female unemployment. In contrast to previous studies, researchers found that an increasing number of unemployed men not only engaged in but also showed pride in accepting some of the more traditional domestic roles like cooking and cleaning [18]. Much like the impact of educational levels discussed earlier, economic shifts may well also impact how gender roles are defined and ultimately how gender impacts risk.

SEXUAL IDENTITY

Thus far this chapter has primarily focused the impact of gender (male and female) on risk and preparedness within communities throughout the globe. It is important to understand that gender and sexuality are not the same thing. For example, one source defined gender as "the socially-constructed roles, behaviors, activities and attributes that a society considers appropriate for a person based on his or her assigned sex at birth" [1]. In other words, if a person was born with the biological component of a female, society will automatically assign traditional female gender roles to that person. These assigned gender roles are based on the cultural norms mentioned earlier and are "socially constructed and...learned through socialization processes" [3].

In Other Words...Defining Gender

The social attributes and opportunities associated with being male and female and the relationships between women and men and girls and boys...these attributes, opportunities and relationships are socially constructed and are learned through socialization processes...In most societies there are differences and inequalities between women and men in responsibilities assigned, activities undertaken, access and control over resources, as well as decision-making opportunities. Gender is part of the broader socio-cultural context.

~United Nations [3]

In contrast to gender, sexuality is limited to a strictly biological assessment and lacks any cultural meaning. This determination of sexuality is determined at birth and primarily relates to the physiological differences between males and females, such as reproductive capacity, body shape, and other distinctive features [19]. However, this definition has not always been sufficient to address the full complexity of how an individual perceives himself or herself and expects to engage within a given society. For example, the World Health Organization (WHO) maintains a working definition of sexuality as a "central aspect of being human throughout life [that] encompasses sex, gender identities and roles, sexual orientation, eroticism, pleasure, intimacy...reproduction...[and] is experienced and expressed in thoughts, fantasies, desires, beliefs, attitudes, values, behaviors, practices, roles and relationships" [20]. This definition is most certainly modern as traditional definitions have primarily understood sexuality as "a 'natural' phenomena [*sic*], intrinsic to an individual's biological constitution" [21].

This modern shift is most evident in the rise of nonheterosexual views such as lesbian, gay, bisexual, and transsexual communities (LGBT). These communities represent individuals who do not see their sexuality (and in some cases gender role) in the biological or traditional roles established earlier. For example, those individuals who identify themselves as gay or lesbian prefer shared sexuality with individuals of the same biological gender (i.e., men with men, women with women). Similarly, those individuals who identify themselves as transsexual or transgender do not identify with the sex into which they are born and ultimately have gender identities "that cross over, move between, or otherwise challenge the socially constructed border between the [biological] genders" [22].

To put this in perspective, it is important to consider some statistics regarding the LGBT communities. According to the Williams Institute at UCLA, approximately 3.5% (nine million total) of American adults consider themselves to be lesbian, gay, or bisexual while an additional 0.9% are self-identified as transgender. Similar studies show these percentages to range from 1.2% to 5.6% of the American population [23]. However, these statistics must be taken somewhat subjectively as they are based on self-reported classifications and do not necessarily take into account continued social or cultural bias toward these sexuality and gender classifications based on religious, cultural, or traditional reasons which occur throughout the world regardless of location.

Because of the cultural stigma of LGBT declarations, human rights groups routinely identify that younger, school-aged children who identify with these sexuality and gender classifications are socially impacted. For example, 42% of surveyed youth who identified themselves as LGBT stated that their communities did not accept their gender identification. Likewise, 92% said they routinely heard negative things about their identification via school, Internet, and peers. Because of these public perceptions, many of the surveyed LGBT youth indicated they were far more honest about themselves and their personal identification in online and social media than in real-world environments [24].

However, much like the process of gender mainstreaming discussed earlier, similar developments have also occurred in relationship to LGBT communities throughout the world. This is most evident in the public and legal acceptance of homosexual marriages (see Figure 5.6). For example, according to the Pew Research Center, public opposition to gay marriage has dropped from 65% in 1996 to less than 40% in 2014 [25]. The support for rights related to same-sex marriage has also increased—particularly in younger generational cohorts. For example, 68% of Millennials and 55% of Generation X favor allowing gays and lesbians to legally marry [25]. Likewise, media representation of the LGBT community is also changing. For example, of the nearly 800 primetime television broadcast series regulars in 2013, 3.3% were presented as LGBT characters. While this number was less than the year before, it has consistently ranged from 2.9% to 4.4%, which roughly matches the general population statistics mentioned earlier for the LGBT community [26].

However, even with the LGBT mainstreaming in some cultures throughout the world, there have been numerous examples of LGBT discrimination in disaster response and recovery. For example, during the response to a 2008 earthquake in Haiti, homosexual men were denied

Figure 5.6 The public acceptance of gay marriage in the United States has increased in recent years. (Source: US Navy.)

food aid because ration schemes were designed for women representing a larger family unit. Likewise, during flooding in Pakistan and Nepal in 2008, individuals identified as transgender were denied entry to aid camp because government-issued identification did not match their physical descriptions [27]. In 2010, after a cyclone in Bangladesh, LGBT people were seen as a "shame on the family" and were excluded from their families to ensure that other family members received a larger portion of disaster relief supplies [28].

In Other Words...Difficulties in Planning for LGBT Communities

Research suggests that development staff often overlook gender identity and sexuality concerns because they cause unease and because of a lack of protocols to deal with these issues across different cultural contexts...Relief efforts typically use the family as a common unit for analyzing and distributing relief services...[which means] relief aid rarely extends to LGBT people...[who] are [already] vulnerable to being forced out of their family living situation as a result of stigma and prejudice.

~Humanitarian Practice Network [27]

This type of challenge was also present after the massive earthquake that struck Japan in 2011 (see Figure 5.7). Because the government of Japan neither prohibits nor protects the rights of self-declared LGBT community members, the Japanese culture tends to push these individuals toward invisibility, marginalization, prejudice, and stigmatization within society. This cultural segregation was certainly magnified during the disaster as gender and sexuality identifications could not be kept private for various response and recovery processes. For example, there was one instance of a transgender individual refraining from using showers in emergency shelters and another individual being openly described as a "cross-dressing deviant fag" by an emergency shelter volunteer [28].

These identified issues for LGBT disaster survivors are not to say that disaster response organizations are not attempting to address these issues. For example, the Salvation Army has made a public effort via traditional and social media to clarify public misconceptions about their willingness to support LGBT community members. The Salvation Army has

Figure 5.7 After the 2011 earthquake in Japan, the LGBT community was disproportionately impacted due to a lack of governmental protection prior to the event. (Source: US Navy.)

streamlined through communications that their service is to provide to "all people…regardless of sexual orientation or any other factor including race, gender, and ethnicity" [29]. Likewise, the American Red Cross has also initiated LGBT specific disaster response funds. For example, after the 2010 earthquake in Haiti, the Atlantic, Olivia, and RSVP cruise lines in conjunction with the American Red Cross created the LGBT Americans for Earthquake Relief in Haiti campaign. This campaign ultimately generated more than 1,300 donations that totaled more than $150,000 [30]. While not specifically generated to support the recovery of the LGBT community in Haiti, the relationship between the American Red Cross and the LGBT community is certainly a progressive step toward direct and indirect acknowledgment of the role of the LGBT community during disaster response and recovery.

Acknowledging these traditional limitations is important for the progression of modern emergency management. At the broadest level, the most effective way for the professional emergency management community to begin to fully understand and address the presence of LGBT constituents and related preparedness challenges is to include LGBT community members in emergency preparedness and planning efforts from the start. This type of approach ensures a clear and consistent perspective approach to the execution of emergency response and recovery as well as a broader integration of LGBT community members in personal preparedness, disaster volunteerism, and donations to support disaster relief. While clearly not all global communities are progressive enough at this point for this level of open involvement, early engagement steps may still be as simple as identifying LGBT communities as vulnerable community sectors in a disaster much like the Philippines did in regard to their response to gender roles during disasters.

Another approach to addressing this issue may simply be broadening the cultural definitions of family structures and relationships. For example, same-sex marriage in the United States (and many other countries) has had increasing legal recognition with more than 200,000 couples in the United States recognized as legally married [31]. However, other countries and cultures do not necessarily project this trend or still prohibit the legal recognition of LGBT relationships. However, it is not necessary for the legal status of a relationship to be determined to receive emergency support in shelters or relief materials. These are broad steps that can begin to embrace these communities in disaster response and relief for strictly humanitarian reasons and without consideration of or influence toward political stances on such issues.

At a more myopic level, the emergency management community will have to implement specific strategies in various response and recovery activities to minimize the continued discrimination of LGBT community members after emergencies and disasters occur. For example, emergency shelter operations are critical disaster recovery processes that serve those displaced or impacted by the given event. However, some very practical issues, such as sleeping, medical support, and bathroom/shower availability, can be sociologically and logistically complicated (see Figure 5.8). For example, gender-based bathroom and shower segregation is traditionally utilized, but can often create stigma and public discomfort for individuals regardless of sexual orientation [32]. While separate facilities could be established for gender-neutral activities, the space and resources to accomplish this are unlikely, given the limits of most disaster relief activities. Likewise, a specific medical program may need to be established to allow for use and storage of specific medications such as hormone treatment or therapies for transgender individuals [31]. Much like the gender-neutral bathrooms, this function may also be difficult or

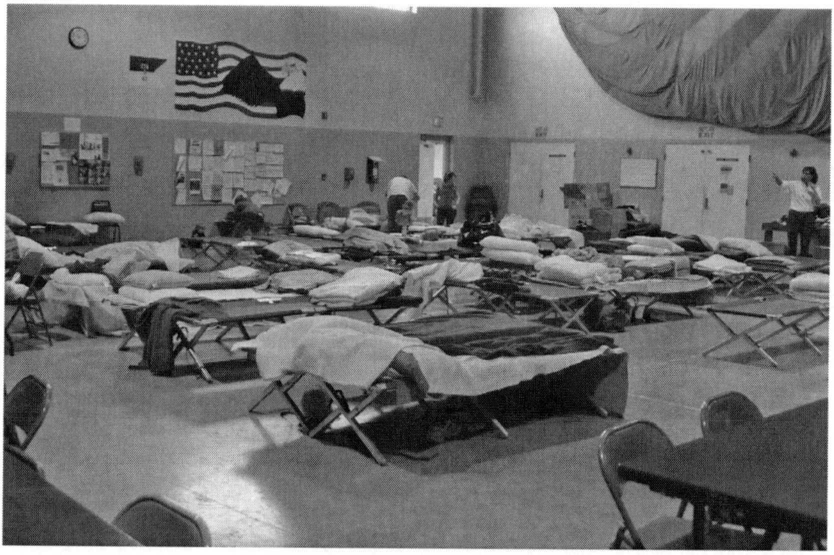

Figure 5.8 Some very practical issues, such as sleeping, medical support, and bathroom/shower availability, can be sociologically and logistically complicated as community leaders attempt to fully integrate impacted citizens of all sexuality classifications. (Source: FEMA/George Armstrong.)

limiting if shelter volunteers lack the understanding and acceptance of these unique medical considerations.

In Other Words...Shelter Challenges for Complex Personal and Family Relationships

There are many circumstances where a strict division between women's bathrooms and men's bathrooms is impractical...[as such] steps should be taken to make it convenient for everyone without compromising the dignity of anyone or forcing someone to publicly reveal their gender choices by requiring them to use a specific toilet or shower facility. A possible solution is to designate a bathroom, especially a single-stalled bathroom if one is available, as being a gender-neutral or family bathroom.

~National Disaster Interfaiths Network [32]

GENERATIONAL DIFFERENCES

The study of generations has a long history in sociological, economical, and cultural contexts. Support for traditional hierarchies structured on the respect and wisdom of aged generations is common in almost all cultures throughout the world. Likewise, industrial countries also place strong generational importance on work status and tenure within organizations and culture. Additionally, there are generational components of life stage (e.g., married, parent, widow, etc.) that also influence how people perceive the world, engage with one another, and ultimately respond during emergencies or disasters. The chronological age, life stage, career stage, and tenure components of the concept of generation will be considered in this component of the chapter to identify current issues and future trends related to challenges and preparedness efforts [33].

The first component of generational status is the consideration of chronological age. Chronological age is often a collective measure for age-related issues such as physical, social, emotional, and cognitive development. These developmental stages are typically divided by arbitrary birth year cut-off dates and divided into generational cohorts that often share societal experiences such as economic circumstances, historical events, etc. that help shape personal and cultural belief systems [33]. There are five cohorts generally utilized to define various age brackets in the United States. For example, those individuals born before 1925 are often referred to as the Greatest Generation or the GI Generation because they lived through the Great Depression and World War II and sought the so-called American dream of college, family, and home [34]. Likewise, those born between 1925 and 1945 may or may not have lived through the Depression, may or may not have been too young to fight in World War II, and ultimately may or may not have been wholly active in Vietnam. They often focused on maintaining the stability created by the previous generation [34]. The next group of people, born between 1946 and 1964, are generally referred to as Baby Boomers or the Sandwich Generation and are generally recognized for a shift back toward cultural elements they experienced as children and then rejected in early adulthood and by the need to simultaneously care for children and aging parents. After the Baby Boomer generation, a group of individuals born between 1965 and 1983 have come to be known as Generation X (or simply Gen X); many grew up as so-called "latchkey kids" and experienced huge progressions in technology and related support systems even though economic independence and viability were not as quickly developed as they were for previous generations. The last defined generational cohort is those individuals born after 1983 that are generally referred to as Millennials. They are digital natives, have broader access to scientific and philosophical information, and are just beginning to establish themselves in the workforce [34].

Interestingly, traditional generational studies would assume that there was a natural and balanced age pyramid where fewer people remained alive and thus part of a particular generational cohort the older the age to which they lived. However, a recent study indicated that global life expectancy had risen by 11 years for men and 12 years for women over the past four decades due to rapid declines in malnutrition and widespread control of infectious diseases [35]. As such, the number of generational cohort members will significantly rise in the future. For example, the United Nations projects that between 2010 and 2050, those individuals over the ages of 65, 85, and 100 will rise by 188%, 351%, and 1004%, respectively [36].

Unfortunately, this shift toward a larger presence of older generations will have significant impacts on emergency and disaster preparedness. For example, a 2009 US Citizens Corps report revealed that adults over the age of 55 were less prepared to cope with disasters. Specifically, 43% of those surveyed over the age of 55 had not taken personal preparedness efforts due to an expected reliance on emergency responders like police, fire, and emergency medical providers compared to younger groups [37]. Moreover, older generations are more likely to depend on maintenance medications or to interact with local health and medical providers, who may be overwhelmed or unavailable during emergencies and disasters (see Figure 5.9).

In Other Words…Age and Disaster Vulnerability

Older adults were the most frequently identified group [in research] among vulnerable populations. These older adults were not classified as vulnerable solely on the basis of age, but as well according to what it means to have special needs. As noted by [researchers]… "age does not make a person vulnerable…rather it is the correlation between advancing age and the likelihood of having special needs that increases frailty."

~Isiah Marshall, Jr. and Shannon Mathews [38]

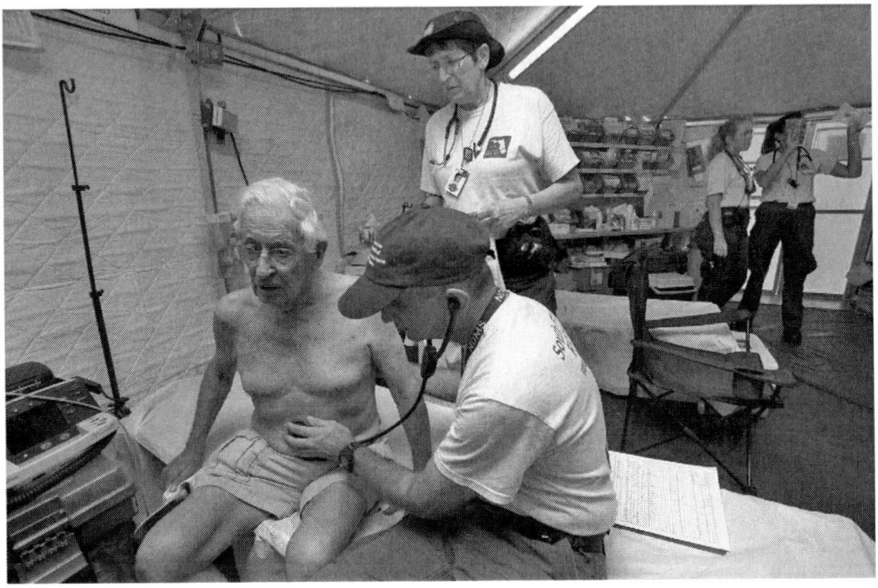

Figure 5.9 Older generations are more likely to be dependent on maintenance medications or to interact with local health and medical providers, who may be overwhelmed or unavailable during emergencies and disasters. (Source: FEMA/ Andrea Booher.)

In addition to the inherent vulnerability of age, older generational cohorts are also more likely to reside in areas where they have lived for long periods of time. For a variety of reasons, including fixed or limited incomes as well as personal comfort in the geographic and physical location, these residences will have a higher likelihood of older infrastructure including safety features and compliance with current building standards and best practices. These personal choices and related generational tendencies can create additional vulnerability during disaster response and recovery. This additional vulnerability was strongly present during the aftermath of Hurricane Katrina, where 75% of the nearly 1,000 deaths were over the age of 75 [38].

Clearly, there is a significant need for emergency management organizations to begin to more strongly consider these various identification roles. While the impact of age and generational roles is more understood, it simply reinforces the need for better education, planning, preparedness, and recovery efforts around how sexuality, gender, and generational roles impact disaster survivors. Hurdles to improving these efforts are common and typically revolve around personal bias, political limitations, funding restrictions, and a targeted focus on issues immediately impacting the community (rather than those that may impact it in the future).

REFERENCES

1. "Disaster Risk Management: A Gender-Sensitive Approach Is a Smart Approach." (n.d.). Brookings Institution. http://www.brookings.edu/~/media/research/files/reports/2013/03/natural%20disasters%20review/nd%20review%20chapter%204. Accessed May 30, 2014.
2. "Social Norms and Cultural Dynamics." (2012). International Cognitive and Culture Institution. http://www.cognitionandculture.net/home/news/58-cfps/2391-social-norms-and-cultural-dynamics. Accessed May 20, 2014.
3. "Making Disaster Risk Reduction Gender Sensitivity." (2009). United Nations. http://www.preventionweb.net/files/9922_MakingDisasterRiskReductionGenderSe.pdf. Accessed June 3, 2014.
4. "Women More Vulnerable to Violence during Flooding." (2008). *The Daily Star.* http://archive.thedailystar.net/newDesign/news-details.php?nid=31477. Accessed June 3, 2014.
5. Wilson, S. et al. (2007). "The Lack of Disaster Preparedness by the Public and Its Effect on Communities." *The Internet Journal of Rescue and Disaster Medicine* 7 (2). http://ispub.com/IJRDM/7/2/11721. Accessed June 3, 2014.
6. "Disasters and Gender Statistics." (n.d.). International Union for the Conservation of Nature. http://cmsdata.iucn.org/downloads/disaster_and_gender_statistics.pdf. Accessed June 5, 2014.
7. "Gender Mainstreaming." (n.d.). UN Women. http://www.un.org/womenwatch/osagi/gendermainstreaming.htm. Accessed June 3, 2014.
8. Mehra, Rekta and Gupta, Getta Rao. (2006). "Gender Mainstreaming—Making It Happen." International Center for Research of Women. http://www.icrw.org/files/publications/Gender-Mainstreaming-Making-It-Happen.pdf. Accessed June 4, 2014.
9. "Gender Equality: Evaluation of Gender Mainstreaming in UNDP." (2006). United Nations Development Program. http://web.undp.org/evaluation/documents/eo_gendermainstreaming.pdf. Accessed June 4, 2014.
10. "Integrating Gender Issues in Disaster Risk Management Policy Development and in Projects." (n.d.). The World Bank. http://www.gfdrr.org/sites/gfdrr.org/files/Guidance_Note_2_Integrating_Gender_Issues_in_Disaster_Risk_Management_Policy_Development_and_in_Projects.pdf. Accessed June 5, 2014.
11. Cauchon, Dennis. (2013). "An American Role-Reversal: Women the New Breadwinners." *USA Today.* http://www.usatoday.com/story/news/nation/2013/03/24/female-breadwinners/2015559/. Accessed June 5, 2014.

12. MacGregor, Donald G., Finucane, Melissa, and Gonzalez-Caban, Armando. (n.d.). "Risk Perception, Adaptation, and Behavior Change: Self-Protection in the Wildland-Urban Interface." *Risk Adaptation and Behavior Change.* http://www.firescience.gov/projects/01-1-7-14/project/01-1-7-14_01-1-7-14_macgregor_finucane_gonzalez-caban_in_press.pdf. Accessed June 5, 2014.
13. Ripley, Amanda. (2008). "Why Men Die More in Floods." *Time.* http://content.time.com/time/nation/article/0,8599,1817603,00.html. Accessed June 5, 2014.
14. "Women, Gender, and Disaster: Men & Masculinities." (n.d.). Gender and Disaster Network. http://www.gdnonline.org/resources/GDN_GenderNote3_Men&Masculinities.pdf. Accessed June 5, 2014.
15. Fothergill, Alice. (1996). "Gender, Risk, and Disaster." *International Journal of Mass Emergencies and Disasters.* http://www.ijmed.org/articles/96/download/. Accessed June 5, 2014.
16. Hillsberg, Alex and Adelman, David. (n.d.). *Finances Online.* http://reviews.financesonline.com/review-of-social-media-and-smartphone-usage/. Accessed June 5, 2014.
17. "Men on Black Sunday." (2013). Women's Health Goulburn North East. http://www.whealth.com.au/documents/work/about-men/FINAL-REPORT-Vol-3.pdf. Accessed June 5, 2014.
18. Pappas, Stephanie. (2011). "'Mancession' Shifts Gender Roles." *LiveScience.* http://www.livescience.com/15695-mancession-recession-shifts-gender-roles.html. Accessed June 6, 2014.
19. "Gender, Households, Community and Disaster Management." (2002). South Pacific Disaster Reduction Program. http://ict.sopac.org/VirLib/TR0282.pdf. Accessed June 6, 2014.
20. "Defining Sexual Health." (2014). World Health Organization (WHO). http://www.who.int/reproductivehealth/topics/sexual_health/sh_definitions/en/. Accessed June 6, 2014.
21. "Sociology and Sexuality." (n.d.). Sydney University. http://www.isis.aust.com/stephan/writings/sexuality/soci.htm. Accessed June 6, 2014.
22. "Transgender vs. Transexual." (n.d.). *Diffen.* http://www.diffen.com/difference/Transgender_vs_Transsexual. Accessed June 8, 2014.
23. Gates, Gary J. (2011). "How Many People Are Lesbian, Gay, Bisexual and Transgender?" Williams Institute. http://williamsinstitute.law.ucla.edu/wp-content/uploads/Gates-How-Many-People-LGBT-Apr-2011.pdf. Accessed June 8, 2014.
24. "Growing Up LGBT in America: View Statistics." (2014). Human Rights Campaign. http://www.hrc.org/youth/view-statistics#.U5UNUPmwJcR. Accessed June 8, 2014.
25. "Gay Marriage." (2014). Pew Research Center. http://www.pewresearch.org/data-trend/domestic-issues/attitudes-on-gay-marriage/. Accessed June 8, 2014.
26. "Where We Are on TV Report 2013." (2013). GLAAD. http://www.glaad.org/whereweareontv13. Accessed June 9, 2014.

27. Knight, Kyle and Sollom, Richard. (2011). "Making Disaster Risk Reduction and Relief Programs LGBT Inclusive." Humanitarian Practice Network (HPN). http://www.odihpn.org/humanitarian-exchange-magazine/issue-55/making-disaster-risk-reduction-and-relief-programmes-lgbtiinclusive-examples-from-nepal. Accessed June 9, 2014.
28. Yamashita, Azusa. (2012). "Beyond Invisibility: Great East Japan Disaster and LGBT in Northeast Japan." Asia-Pacific Human Rights Information Center. http://www.hurights.or.jp/archives/focus/section2/2012/09/beyond-invisibility-great-east-japan-disaster-and-lgbt-in-northeast-japan.html. Accessed June 9, 2014.
29. "Debunking the Myth of LGBT Discrimination." (2014). The Salvation Army. http://www.salvationarmyusa.org/usn/nodiscrimination. Accessed June 10, 2014.
30. "Lesbian and Gay Haiti Fund Tops the List of Donors to American Red Cross." (2010). PR Newswire. http://www.prnewswire.com/news-releases/lesbian-and-gay-haiti-fund-tops-the-list-of-donors-to-american-red-cross-82885087.html. Accessed June 10, 2014.
31. "Working with Lesbian, Gay, Bisexual, and Transgender Community: A Cultural Competency Guide for Emergency Responders and Volunteers." (n.d). Human Rights Campaign. http://www.hrc.org/files/assets/resources/EmergencyResponders_-_LGBT_Competency.pdf. Accessed June 11, 2014.
32. "LGBT Needs & Disaster." (2008). National Disaster Interfaiths Network. http://www.n-din.org/ndin_resources/tipsheets_v1208/25_NDIN_TS_LGBTNeeds.pdf. Accessed June 11, 2014.
33. "Age and Generations." (2014). The Sloan Center on Aging and Work. http://www.bc.edu/research/agingandwork/projects/generations.html. Accessed June 12, 2014.
34. "Intergenerational Considerations: Generalized Characteristics of Generational Cohorts." (n.d.). United Universalist Association. http://www.uua.org/documents/hoertdoerferpat/070622_intergencharacter.pdf. Accessed June 12, 2014.
35. Boseley, Sarah. (2012). "Life Expectancy around the World Shows Dramatic Rise, Study Reveals." *The Guardian*. http://www.theguardian.com/society/2012/dec/13/life-expectancy-world-rise. Accessed June 13, 2014.
36. "Living Longer." (2014). National Institute on Aging. http://www.nia.nih.gov/research/publication/global-health-and-aging/living-longer. Accessed June 13, 2014.
37. "Disaster Preparedness for the Vulnerable." (2012). National Association of Triads. http://www.nationaltriad.org/tools/alerts/Disaster%20Preparedness%20for%20the%20Vulnerable.pdf. Accessed June 14, 2014.
38. Marshall, Isiah Jr. and Mathews, Shannon. (n.d.). *The Journal of Aging in Emerging Economics*. http://www2.kent.edu/sociology/resources/jaee/upload/article_2.pdf. Accessed June 14, 2014.

6

Self-Reliant Communities

The internet can be a powerful enabling technology fostering the development of communities because it supports the very thing that creates a community—human interaction.

~Esther Dyson
Release 2.0: A Design for Living in the Digital Age [1]

Publicness challenges the notion of the stranger.

~Jeff Jarvis
Public Parts: How Sharing in the Digital Age Improves How We Work [2]

COMMUNITIES AND CULTURE

Previous chapters discussed the impact of cultural norms on choices and ultimately risk from emergencies and disasters. This chapter will begin the consideration of how communities—both traditional and virtual—impact how people engage and participate in disaster processes that range from personal preparedness and risk reduction to large-scale needs from government and other industry providers. This is particularly relevant as the scope and behaviors of communities are changing from geographically based to virtual based. For example, an individual with a unique passion for a particular hobby (e.g., underwater basket weaving) might be in the extreme minority in one limited geographic area (e.g., home city), but when connected into a broader, virtual community, the presence and connectivity around this particular hobby would be magnified greatly.

To understand this potential impact, the difference between culture and community must be understood. Culture is often defined as a set of rules, perceptions, language, history, and other like elements which affect behavior and attitude for a given set of people. Moreover, culture is often defined or embodied in books, songs, literature, and, more modernly, blogs and websites [1]. Culture is emulated or copied in future generations seeking similar behaviors. For example, wardrobe features of the 1980s culture (e.g., side pony tail hair designs and leggings) are becoming increasingly popular again. In contrast, a community is based on a set of relationships, which cannot be copied as they are unique to a given set of circumstances and experiences. For the vast majority of history, these communities have been built around geographic definitions or boundaries often tied to sociopolitical or, in some cases, religious collections (see Figure 6.1). However, with the rise of social-media-based digital technologies and a so-called shrinking world, the definition of community has broadened and now an individual can be a part of a multitude of communities based on traditional geography as well as ethnicity, religion, interests, age, gender, sexuality, or many other social factors [3].

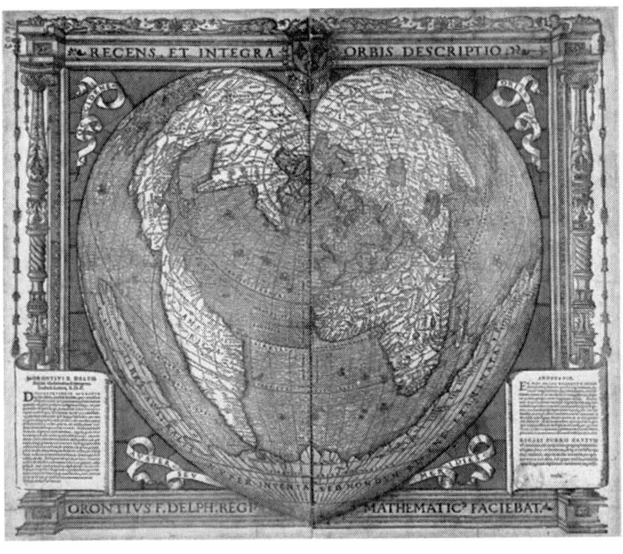

Figure 6.1 For the vast majority of history, communities have been built around geographic definitions or boundaries with associated sociopolitical or religious connections. (Source: Oronce Fine.)

Understanding both traditional and modern views of community is particularly important for modern emergency management, homeland security, and public safety officials seeking to understand how communities will respond before, during, and after a major disaster. One of the fundamental rules of emergency management has been the concept that all disasters are local. This rule implies that a geographically based community that includes the area impacted by the event must ultimately be responsible for response and recovery. While natural hazards still impact a given geographic area, response and recovery support can now come from virtual and traditional sources simultaneously. In some examples, nongeographic emergent volunteerism groups have developed after events and have shared virtual "ownership" of the recovery in significant and oftentimes more impactful ways than local government leaders [4]. Likewise, digital volunteerism groups like Translators without Borders, Humanity Road, and NetHope have done tremendous work providing digital support for locally impacted events throughout the world [5] (see Figure 6.2).

Traditional communities based around geographic boundaries have not always had connectivity to the representative government. Oftentimes over the course of history, community members have simply lived in relationship with one another with assistance and support provided to one another and most often outside any government intervention. However, with the start of

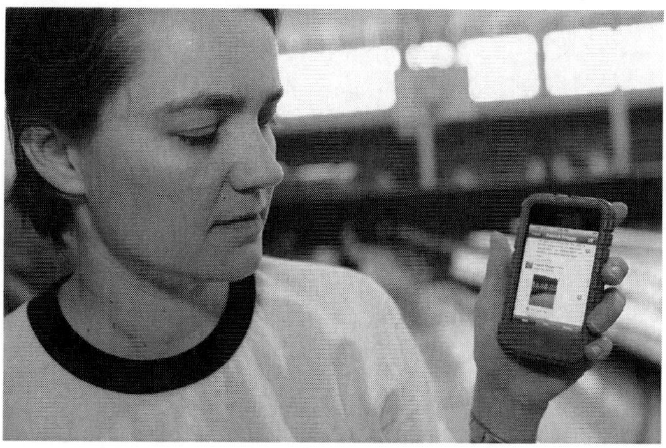

Figure 6.2 Digital volunteerism groups have provided tremendous work in disaster recovery by providing digital support for locally impacted events throughout the world. (Source: FEMA/David Fine.)

Figure 6.3 With the start of the Great Depression in 1929 and the rise of Roosevelt's New Deal recovery programs, there was a significant shift toward a much greater dependency on governmental support. (Source: National Archives and Records Administration.)

the Great Depression in 1929 and the rise of Roosevelt's New Deal recovery programs, there was a significant shift toward a much greater dependency on governmental support (see Figure 6.3). This increase in government intervention creates an unstable balance between communities and their government. For example, the level of engagement in private schools (with limited government involvement) from parents, teachers, students, and supports is significantly higher than in public schools, which are government funded, supported, and controlled.

In Other Words...The Impact of Government Intervention on Communities

Government can play a divisive role vis-a-vis communities. Often, the more government provides, the less community members themselves contribute.

~Esther Dyson [1]

The level of engagement of government at local, regional, state, and federal levels has increased since the rise of the New Deal. For example, as a measure of government growth, Standard & Poor's has stated that current government spending has increased by 800% since the 1940s [6]. Moreover, this level of government engagement in local communities has been paralleled in disaster and emergency management. For example, the Stafford Act of 1988 (later amended in 2007) established six core statutory responsibilities for federal engagement and support of disaster preparedness and response which ultimately broadened "the scope of existing disaster relief programs" [7]. Two of the most significant components of the Stafford Act (and corresponding federal programming) are individual assistance (IA) and public assistance (PA), which provide financial support for governmental organizations and individuals who are impacted during emergencies and disasters. Likewise, Congress passed the Disaster Mitigation Act in 2000, which expanded "federal engagement with a mandate to help state and local governments develop and carry out hazard mitigation measures" by providing no less than 75% reimbursement for eligible government expenditures during disaster response and recovery [7]. Unfortunately, these types of legislative acts do provide support, but ultimately create an expectation and dependence on government rather than a balance of individualized community preparedness and risk reduction measures.

In Other Words...The Role of Responsibility in Disaster Response and Recovery

In conformity with the principles of federalism, [federal] statutes affirm the federal government's role of "assisting" states and local governments in their disaster management responsibilities.... however, not of a supportive kind. Individuals, states, and local governments are not the main bearers of responsibility. Under these statutes, "assistance" means that the federal government takes primary responsibility...[which] also manifests itself firmly in state budgets...[where] more than 80% of required funding for emergency and disaster activities is covered by federal funds.

~Amy Crabill and Yvonne Rademacher [7]

This imbalance in disaster preparedness and risk resiliency between individual communities and government peaked during response to Hurricane Katrina in 2005. For example, a 2006 report released by the Select Congressional Bipartisan Committee to Investigate the Preparation for and Response to Hurricane Katrina identified over 90 findings of failure at all levels of government regarding the coordination of response and recovery to the hurricane [8,9] (see Figure 6.4). While these findings were likely accurate deficiencies in the disaster management system utilized to respond to the impacts of the disaster, the report also identified "exceptions to the rule" of individuals and local communities who took personal actions (e.g., evacuating local nursing homes) that ultimately reduced risk [9]. Likewise, the report also identified the personal sense of abandonment that many of the local community felt because of the lack of coordinated support from government. For example, one Hurricane Katrina

Figure 6.4 The Select US Congressional Bipartisan Committee to Investigate the Preparation for and Response to Hurricane Katrina found over 90 findings of failure at all levels of government regarding the coordination of response and recovery. (Source: Library of Congress.)

survivor said, "We were abandoned. City officials did nothing to protect us.... When you feel like this you do one of two things, you either give up or go into survival mode. We chose the latter" [9]. This attitude was reflected in a Knight–Ridder survey that found Katrina survivors were not necessarily of a particular ethnicity, race, culture, or socioeconomic strata, but rather impacted by their belief system about disasters (and how to recover) before the event struck [10]. The independent community that had previously relied on each other had completely eroded with near full dependence on government for basic and fundamental needs during response and recovery.

DIGITAL COMMUNITIES

Coincidentally, at nearly the same time as Hurricane Katrina, social media sites were beginning to gain a foothold online and in certain communities. For example, Facebook started in 2004 as a social network available only to Harvard University students, but quickly expanded to other higher education institutions throughout the country. Likewise, Twitter was started in 2006 by a couple of ex-Google employees as an attempt to create community through short messaging. Interestingly, in both cases these technological forms originated out of traditional community environments—college for Facebook and employment for Twitter. However, over the next 5 years the number of users increased and the foundations of digital community were created. For example, it took Twitter slightly more than 3 years to reach its billionth tweet [11]. Likewise, Facebook went from one million users in 2004 to 58 million users in 2007 to 608 million users in 2010 [12].

Within 5 years of Hurricane Katrina, the size and scope of digital communities made a significant impact on another large-scale catastrophic event. Specifically, on January 12, 2010, a 7.0 magnitude earthquake struck Haiti not far from the capital city of Port-au-Prince. Due to the proximity of a high-population area, there were an estimated three million people impacted as well as a "serious loss of life" [13,14]. Given the lack of highly developed infrastructure in Haiti and the overall high rate of poverty, Haiti was already highly vulnerable even before the catastrophic earthquake. For example, the World Bank estimated that more than half of Haiti's population lived on less than $1 per day, 80% lived on less than $2 per day, and the unemployment rate was more than 40% [15]. Because of the high risk and immediate impact to life from the earthquake, disaster response and humanitarian aid quickly flowed into the country from all parts of the world.

However, in addition to the high levels of poverty and poor infrastructure in Haiti, emergency responders and aid groups were faced with local communication challenges. Specifically, the primary spoken language in Haiti was Creole with minor pockets (no more than 10%) of French. Moreover, local culture viewed the spoken language of Creole as a symbol of national identity in Haiti [16]. Many local issues, reports, and situational awareness were initially limited due to the lack of translators who could speak the language and provide the needed information to the various support teams. Unfortunately, traditional solutions for providing translators with various response teams were extremely limited as there were only approximately 400,000 individuals (mostly in the United States) who spoke Haitian Creole [17].

The solution to this problem was the utilization of digital technologies and the leveraging of virtual communities. These virtual communities simultaneously leveraged local community characteristics (e.g., language) and created a non-geographic-dependent community designed to help those in actual need. Specifically, in the case of the response to the Haiti earthquake, an emergent group of volunteers from more than 50 countries worldwide created an "online translation and information processing service that connected the Haitian people with each other and with the international aid efforts" [18]. This emergent group—called Mission 4636—leveraged an open source crisis mapping system called Ushahidi and the local text messaging infrastructure, which surprisingly was only minimally impacted during the earthquake. Specifically, a local Haitian with disaster-related needs would send a text message to the identified number, which would then be translated by Haitian-speaking volunteers throughout the world and converted into usable data including text and geospatial data. This information was then routed to appropriate international emergency response organizations to more effectively and efficiently address the issue. This process was able to manage more than 80,000 text messages sent in the first 30 days of recovery. Thus, "this coordinate[d] response was able to meet affected Haitians' urgent needs for medical care, food, water, security, and shelter in real time...and filter those requests...by geographic area or type of need" [19].

The impact of virtual connectivity was seen again in April 2011 in the state of Alabama after more than 60 tornadoes struck various parts of the state over a 3-day period of time. As part of a large outbreak of tornadoes and on the heels of continued recovery from the BP oil spill the previous year, the state of Alabama was overwhelmed as physical communities throughout the state were catastrophically impacted and in some cases completely destroyed. In the days and weeks following the event, local and state emergency management organizations attempted to address local and regional needs. Unfortunately, given the scope and complexity of the event, there were often areas where resources were not delivered in a timely manner. This meant that basic needs—particularly for vulnerable populations—were often unavailable for extended periods of time. Consequently, this deficiency left many disaster survivors seeking other solutions, often via social networks like Facebook and Twitter.

For example, a group of individuals in an unimpacted area of Alabama created an emergent volunteerism group called Toomers for Tuscaloosa. (Note: the name is in reference to local landmarks associated with two local universities.) Within weeks, Toomers for Tuscaloosa had set up not only donations receiving sites and systems, but also a strong social media presence through the leveraging of hashtags to address localized needs (#alneed) and support (#alhave) (see Figure 6.5). This social presence translated to

Figure 6.5 Emergency volunteerism groups like Toomers for Tuscaloosa have provided significant support online and through the collection of donated goods to support disaster recovery. (Source: FEMA/Steve Zumwalt.)

more than 82,000 Facebook and nearly 3,000 Twitter followers, which was 27 times greater than the state emergency management agency had at the time [4]. Clearly, the general public—and related physical communities—were significantly improved by the presence and capability of the virtual communities, which spread not only through one geopolitical state, but also often throughout the world.

Nearly a month later the same type of virtual-community-supported, communitywide response to the EF-4 tornado that struck Joplin, Missouri, and killed 116 people took place [20]. Specifically, a woman in a nearby community created a Facebook group called "Joplin Tornado Info" that was initially built to help family members find each other, but quickly turned into a place where disaster-related needs could be shared and addressed in a fashion similar to the Toomers for Tuscaloosa group. Within the first 24 hours, there were 30,000 followers and more than 80 million page views were eventually generated [21]. Just like the Alabama emergent group, the Joplin group quickly filled the void in the physical community due to the significant impact to local infrastructure and the disruption of local government and emergency services. The tornado page ultimately was considered by online social media news outlet Mashable for its annual Best Social Good Award [21].

The Toomers for Tuscaloosa and Joplin Tornado Info Facebook pages occurred independently of one another in two divergent physical communities. The independent creation of each group of emergent volunteers shows strong anecdotal evidence that human behavior via social media is organic, beneficial, and ultimately completely disconnected from traditional forms of community. Due to the incredible power of social media systems and the ubiquitous availability of digital and mobile technologies, the concept of connected disconnectedness will continue for disaster response and recovery throughout the world.

IMPACT OF TECHNOLOGY ON GLOBAL COMMUNITIES

The use of social media outlets like Facebook and Twitter during emergencies and disasters to leverage virtual communities is only fully possible if technology is widespread and broadly available to users throughout the world. This is most commonly evident through technology developments related to mobile phones and high-speed and reliable access to Internet networks. With only 41% of the world's households having a connection to the Internet and more than 1.1 billion households in the developing world not connected, this needed access is not a foregone conclusion throughout the world [22]. Likewise, only 61.1% of the world has access to mobile phones to access and engage in instantaneous information [23]. Understanding the current access and considering future trends will provide further clarity regarding the rise of independent and virtual global communities.

Both of these access issues must be considered further to fully grasp the impact on communities at large. For example, more men than women use the Internet globally, although the difference between the two is not significant. Specifically, 37% of women (1.3 billion) compared to 41% of men (1.5 billion) have some level of access to the Internet. Of those users, 475 million are female Internet users from developed countries as well as an additional 483 million male users. The gender gap is more pronounced in the developing world, where 16% fewer women than men use the Internet [22]. Likewise, Europe (77%) and Africa (7%) represent the continents with the highest and lowest levels of Internet usage, respectively. However, Internet penetration usage rose the fastest in African between 2009 and 2013 as users increased by 27% [22].

Likewise, the global use of smartphones has also grown in unique ways. Specifically, the use of smartphones surpassed one billion in 2012 and rose to 1.75 billion in 2014 with a continued fast-paced trajectory through 2017 [23]. Moreover, nearly 40% of all global mobile phone users used a smartphone at least monthly in 2014; however, this number will exceed 50% by 2017. This growth is related to an increasingly affordable network presence throughout the world and a decreasing cost of production for the individual devices. Much like Internet access, the use and availability of mobile phones are widespread in Europe and North America, but much less likely in low income areas, particularly in Latin America, Middle East, Africa, and much of Southeast Asia [23].

The fascinating issue is where these two technologies intersect. Specifically, nearly 50% of mobile phone users (more than two billion people) accessed the Internet from their mobile phones in 2014. The availability of expanding cellular and data networks in conjunction with the availability of smartphones is contributing to this quick rise and is projected to show double-digit growth in mobile phone Internet access over the next 5 years [23]. This rise is particularly true in the United States where (as of 2013) 63% of adult mobile phone users accessed the Internet from their mobile phone and 34% accessed the Internet almost solely from the phone rather than some combination of other access points (e.g., laptops) [24]. This represents a 100% increase since 2009 and a steady and consistent growth over that period that shows no indication of slowing down or leveling off [25]. Likewise, the trend of using mobile phone Internet is even greater in younger generations and future cell phone users as 85% of cell phone users aged 18–29 and 73% of those aged 30–49 actively use their cell phones to access the Internet [25] (see Figure 6.6).

The second significant trend related to the merging of Internet usage and the increasing presence of smartphones is related to its relationship to demographic changes in developed countries like the United States. Specifically, the United States is experiencing a significant shift in its racial diversity with significant increases in the Hispanic and African American populations, which are projected to reach 29% and 13%, respectively [26]. Interestingly, 68% of all Hispanic mobile phone users actively utilize their devices to access the Internet, as do 74% of all African American users [25]. These types of trends are important to understand and monitor into the future as they may ultimately impact how people send and receive information within communities and certainly how disaster information is processed.

On the other end of the global socioeconomic spectrum, many developing countries have approached telephony and Internet access differently. For example, in developing countries in Africa and parts of

Figure 6.6 The use of cell phones to access the Internet is greatest among younger generations and will become a critical disaster response and recovery tool. (Source: FEMA/Marvin Nauman.)

Southeast Asia, more than 50% of Internet users do not access Internet on their personal computers. Some studies have shown that this figure is as high as 70% in Egypt [27]. Moreover, mobile phone subscriptions in sub-Saharan Africa have risen from 16 million in 2000 to 376 million in 2008. This utilization of mobile phones has improved local economies, increased jobs, and ultimately improved safety and preparedness from various natural and human-caused disasters [27]. In contrast, there are fewer than three landlines per 100 local citizens there. The service provision on those landlines is poor as some areas have reported 36 days of service interruption over an average year with individual disruptions lasting an average of 37 hours. Moreover, many areas have significant delays (as many as 100 days) to facilitate a connection with a communication company via these landlines, which sometimes requires bribes or other incentives [28]. Thus, the direct and indirect impacts of landline infrastructure and service have quickly shifted users in developing countries away from traditional infrastructure and toward mobile devices for both communication and Internet access.

In Other Words...Mobile Phone Infrastructure throughout the World

While the telecommunications industry in the United States, Canada and Europe invested in landlines before moving to mobile phone networks, the mobile phone has effectively leapfrogged the landline in Africa. After all, landlines require that wires be installed on every road and into every community, with smaller lines into every household. This can be prohibitively expensive, especially in countries with poor roads, vast distances and low population densities. Mobile phone coverage in sub-Saharan Africa, by contrast, is primarily provided via a network of specialized base stations, which can provide service to a 5–10 kilometer radius. Due to unreliable electricity supplies across Africa, the base stations are primarily powered by diesel generators.

~Jenny C. Aker and Isaac M. Mbiti [29]

This widespread access to the Internet is being addressed in a variety of ways. For example, since 2013, Google has supported Project Loon, which uses balloons to create a wireless Internet network in remote areas with limited infrastructure. Specifically, large balloons are floated in the stratosphere (twice as high as airplanes fly) where they can leverage

Figure 6.7 Google has supported Project Loon, which uses balloons to create a wireless Internet network in remote areas with limited infrastructure. (Source: Julian Colton.)

the many layers of wind to directionally move, rise, or fall to maximize the connectedness of the balloon network. People access the balloon network by attaching a special Internet antenna to their homes or residences to allow a signal to bounce up to the balloon network and back down to the global terrestrial Internet. This concept was first piloted in New Zealand and has been tested in other parts of the world including parts of California's Central Valley [30]. While results are still being collected, Google's Loon project represents one of those opportunities to bring Internet to broader audiences and ultimately improve the virtual community capabilities that could exist throughout the world (see Figure 6.7).

In Other Words...The Current Capabilities of Google's Balloon Project

The Loon balloons are one of Google's "moonshot" projects and are a part of Google's somewhat mysterious "X" research division. Google hopes that the balloons will be able to provide internet connectivity for people around the world who currently don't have access. After one year, the balloons have been rapidly improving…[as] the balloons are delivering 10× more bandwidth, 10× steer-ability, and are staying up 10× as long.

~Ben Zigterman [31]

On the other end of the spectrum, Google has another project that is significantly impacting the physical community and its capability to extend into virtual areas. Specifically, Google Fiber is an Internet and television service created by Google that would provide supercharged speed over the service. More interestingly, the basic Google Fiber service (average Internet speed) is being provided for free, which is minimally $40 or less than the charges from other standard Internet service providers (ISP) [32]. While this service may ultimately be revolutionary, it is currently only available in Kansas City, Missouri; Provo, Utah; and Austin, Texas. However, Google is scheduling expansion to nine additional cities, including Atlanta, Georgia, and Portland, Oregon, within the next few years [33]. In areas like Kansas City, this access to quick, reliable, and inexpensive Internet is being leveraged to create communities around entrepreneurial and "hacker" lifestyles. For example, the Kansas City Startup Village (KCSV) utilizes older homes in proximity to each other with as many as four Internet access points in each room (other than the bathroom) to encourage a blending of physical and virtual community networks [34].

SPONTANEOUS COMMUNITY

In many ways, access to the Internet and by extension social media systems has radically changed how people access information and engage in relationships or community with one another. For example, many people "now take…for granted that any piece of information we want is likely a search away… [but] we are coming to rely on the idea that people we want to meet are just a connection away" [2]. Within social media systems, this one-to-one interaction can often be spontaneous or serendipitous due to the (previously unknown) shared interest in a real community or a virtual representation of that community. This is particularly critical in emergencies and disasters as this spontaneity creates "social glue" and can create a transformative presence between the impacted individuals simply through the ability to virtually connect in a quick and genuine way [35].

This concept has already been considered for emergent volunteerism groups like Toomers for Tuscaloosa and the Joplin Tornado Info Facebook groups; however, there is an additional extension that must be considered further. Specifically, the capability for the crowd to collectively search as a research and/or investigative component is real, quick, and powerful. Numerous law enforcement agencies throughout the world have begun using social media to collect and aggregate data related to an individual or string of local community crimes. However, this concept goes much farther when the moral and philosophical scope of the crime or act of terrorism goes beyond the local physical community directly impacted.

For example, at the conclusion of game 7 in the 2011 Stanley Cup hockey finals, when the Boston Bruins defeated the Vancouver Canucks in Vancouver, a riot started where "some bold troublemakers" threw bricks through windows, tipped portable toilets, and set fires throughout the central part of Vancouver [36]. Interestingly, within hours of the beginning of the riots, the members of the local community immediately started identifying those who participated in the riots and started online social media and blog sites to gather photos, videos, and other identification material related to the riots [37]. This flow of information via online sources continued as the Vancouver Police Department (VPD) received 53 e-mails with attached videos, 676 YouTube links, 708 images via e-mail, and more than 1,011 hyperlinks to other social media sites like Facebook and Twitter as well as nearly 300 tips to the local crime stoppers' phone line. The VPD quickly set up a separate site to collect and aggregate the photos, including a top 10 list of unidentified criminals related to the riot [37]. Interestingly, this site is still leveraged by VPD and has received more than two million Internet visitors who have looked at the posted riot images nearly 16 million times, which has ultimately led to nearly 300 related arrests [38]. This type of social-media-based crowd investigation clearly added to the capabilities of the traditional investigation's effectiveness.

In Other Words...The Impact of Mobile Phones on Criminal Observation during the 2011 Stanley Cup Riots

Some seemed to revel in the rampage, recording the vandalism on cell phones and video cameras. A few congratulated those who tried to attack police and others erupted with cheers every time something was damaged. Fans...sang a drunken tune as they danced on an overturned vehicle...Fans wandered amid the chaos, some with bandanas or T-shirts pulled over their faces—either to hide their faces from police and cameras or to guard against the smoke.

~Associated Press [36]

Unfortunately, the positive benefit of the large-scale virtual crowd investigations is not always as beneficial as the response to the Vancouver riots of 2011. Specifically, in response to the Boston Marathon bombing in April 2013, the Federal Bureau of Investigation (FBI) distributed information to the general public about some of the evidence found in the debris of the explosion, including multiple pictures of the bomb materials and eventually of the two suspects (see Figure 6.8). The social-media-based crowd (primarily using information collection sites like Reddit) quickly identified high-resolution photos of the two terrorists that were ultimately clearer than those released by law enforcement. Unfortunately, the quick and amplifying process of social media also unintentionally circulated misinformation. For example, Twitter users

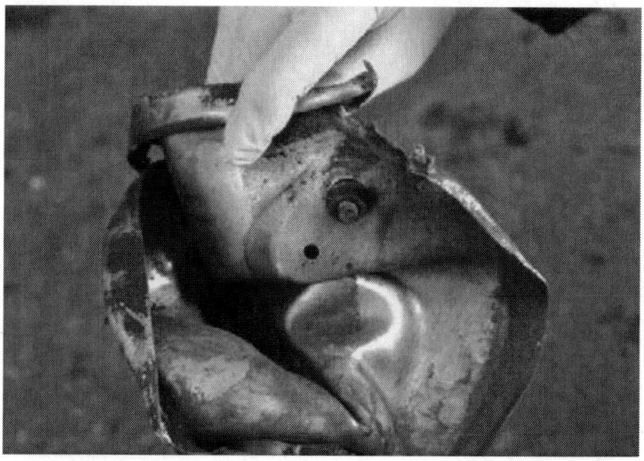

Figure 6.8 The US Federal Bureau of Investigation (FBI) distributed information to the general public about some of the evidence found in the debris of the explosion during the Boston Marathon including multiple pictures of the bomb materials and, eventually, of the two suspects. (Source: FBI.)

quickly spread information that a missing Indian American student was one of the bombers even though public pictures shared by the FBI did not match his description. This "wildfire of speculation" was only stopped after NBC News and other major media outlets began to publicly contradict the social media reports [39].

In Other Words...The "Dangerous Experiment" of Crowd Investigations

The crowdsourcing part of wanted posters is about making sure as many people as possible see the picture. It is emphatically not about making sure any allegations resulting from the picture are made public…[but social investigations] aren't crowdsourcing, it's just speculating; there is little advantage in getting the crowd involved…and the major downside [is] that someone's life might be ruined based on who they [might] look like.

~Alex Hern [39]

In the end, the impact of communities and crowds—whether physical or virtual—is having a significant impact on emergency management and public safety issues. Given the complexity of the changing face of communities and the continued forecasted growth and integration of social media, there is no reason to project that social-media-based emergent behavior after emergencies and disasters of all sizes and scope will continue into the future. So the major issue for the profession of emergency and crisis management is how to embrace these behaviors and leverage them for the overall ability of the physical community.

REFERENCES

1. Dyson, Esther. (1997). *Release 2.0: A Design for Living in the Digital Age.* New York: Broadway.
2. Jarvis, Jeff. (2011). *Public Parts: How Sharing in the Digital Age Improves How We Work.* New York: Simon and Schuster.
3. Niven, Rosie. (2013). "The Complexity of Defining Community." *The Guardian.* http://www.theguardian.com/voluntary-sector-network/2013/may/03/community-spurs-fans. Accessed June 16, 2014.
4. Crowe, Adam. (2012). *Disasters 2.0: The Impact of Social Media on Modern Emergency Management.* Boca Raton, FL: CRC Press.

5. Crowe, Adam. (2013). *Leadership in the Open: A New Paradigm for Emergency Management*. Boca Raton, FL: CRC Press.
6. "Government Spending." (n.d.). New Braunfels Independent School District. Chapter 10. http://www.nbisd.org/users/0022/docs/Economics/chap10.pdf. Accessed June 16, 2014.
7. Crabill, Amy and Rademacher, Yvonne. (2012). "Breaking the Cycle of Reliance on Federal Help after Disasters." *Emergency Management Magazine*. http://www.emergencymgmt.com/disaster/Breaking-Reliance-Federal-Help-After-Disasters.html. Accessed June 17, 2014.
8. Hsu, Spencer S. (2006). "Katrina Report Spreads Blame." *The Washington Post*. http://www.washingtonpost.com/wp-dyn/content/article/2006/02/11/AR2006021101409.html. Accessed June 17, 2014.
9. "A Failure of Initiative." (2006). Final Report of the Select Bipartisan Committee to Investigate the Preparation for and Response to Hurricane Katrina. 109th Congress. http://orise.orau.gov/csepp/documents/planning/evacuation-documents/federal-reports/mainreport.pdf. Accessed June 17, 2014.
10. Ripley, Amanda. (2009). *The Unthinkable: Who Survives When Disaster Strikes and Why*. New York: Harmony.
11. "#Numbers." (2011). Twitter Blog. https://blog.twitter.com/2011/numbers. Accessed June 17, 2014.
12. Sedghi, Ami. (2014). "Facebook: 10 Years of Social Networking, in Numbers." *The Guardian*. http://www.theguardian.com/news/datablog/2014/feb/04/facebook-in-numbers-statistics. Accessed June 17, 2014.
13. "Red Cross: 3M Haitians Affected by Quake." (2010). *CBS News*. http://www.cbsnews.com/news/red-cross-3m-haitians-affected-by-quake. Accessed June 18, 2014.
14. "7.0 Quake Hits Haiti: 'Serious Loss of Life' Expected." (2010). *CNN World*. http://www.cnn.com/2010/WORLD/americas/01/12/haiti.earthquake. Accessed June 18, 2014.
15. Sauter, Michael B., Hess, Alexander E. M., and Weigley, Samuel. (2012). "The 10 Poorest Countries in the World." *Fox Business*. http://www.foxbusiness.com/markets/2012/09/14/10-poorest-countries-in-world. Accessed June 18, 2014.
16. Valdman, Albert. (n.d.). "Creole: The National Language of Haiti." Indiana University. http://www.indiana.edu/~creole/creolenatllangofhaiti.html. Accessed June 19, 2014.
17. Thompson, Irene. (2013). "Creole Languages." All World Languages. http://aboutworldlanguages.com/creole-languages. Accessed June 19, 2014.
18. "Mission 4636." (n.d.). Mission 4636. http://www.mission4636.org. Accessed June 19, 2014.
19. Spivack, Emily. (2011). "A Look Back at Haiti and Mission 4636 with Josh Nesbit and Patrick Meier." *PopTech*. http://poptech.org/blog/a_look_back_at_haiti_and_mission_4636_with_josh_nesbit_and_patrick_meier. Accessed June 19, 2014.

20. Esposito, Richard. (2011). "Joplin Death Toll at 116 Making It Deadliest Tornado in Nearly 60 Years" *ABC News*. http://abcnews.go.com/US/joplin-tornado-death-toll-116-makes-deadliest-single/story?id=13662193. Accessed June 20, 2014.
21. Henry, Lindsay. (2011). "Joplin Tornado Info Facebook Page Up for Mashable Award." KOAM. http://www.koamtv.com/story/16139615/joplin-tornado-info-facebook-page-up-for-award-from-mashablecom. Accessed June 20, 2014.
22. "The World in 2013: ICT Facts and Figures." (2013). ITU. http://www.itu.int/en/ITU-D/Statistics/Documents/facts/ICTFactsFigures2013-e.pdf. Accessed June 20, 2014.
23. "Smartphone Users Worldwide Will Total 1.75 Billion in 2014." (2014). eMarketer. http://www.emarketer.com/Article/Smartphone-Users-Worldwide-Will-Total-175-Billion-2014/1010536. Accessed June 20, 2014.
24. "Mobile Technology Fact Sheet." (2013). Pew Internet Research Project. http://www.pewinternet.org/fact-sheets/mobile-technology-fact-sheet. Accessed June 21, 2014.
25. "Cell Internet Use 2013." (2013). Pew Internet Research Project. http://www.pewinternet.org/2013/09/16/cell-internet-use-2013. Accessed June 21, 2014.
26. Wade, Lisa. (2012). "U.S. Racial/Ethnic Demographics: 1960, Today, and 2050." *Sociological Images*. http://thesocietypages.org/socimages/2012/11/14/u-s-racialethnic-demographics-1960-today-and-2050. Accessed June 22, 2014.
27. Munford, Monty. (2010). "Developing (and Developed) Countries Embrace the Mobile Internet." *The Telegraph*. http://www.telegraph.co.uk/technology/news/8220274/Developing-and-developed-countries-embrace-the-mobile-internet.html. Accessed June 22, 2014.
28. Aker, Jenny C. and Mbiti, Isaac M. (2010). "Mobile Phones and Economic Development in Africa." *Journal of Economic Perspectives*. http://sites.tufts.edu/jennyaker/files/2010/09/aker_mobileafrica.pdf. Accessed June 23, 2014.
29. Aker, Jenny C. and Mbiti, Isaac M. (n.d.). "Mobile Phones and Economic Development in Africa." University of Berkeley. http://businessinnovation.berkeley.edu/Mobile_Impact/Aker-Mbiti_mobile_phones_Africa.pdf. Accessed June 22, 2014.
30. "Loon for All." (2014). Google Project Loon. http://www.google.com/loon. Accessed June 23, 2014.
31. Zigterman, Ben. (2014). "Google's Internet Balloons Cause Panic in Yet Another Country." BGR. http://bgr.com/2014/06/20/google-loon-balloon-internet-project. Accessed June 23, 2014.
32. Lloyd, Craig. (2013). "What Is Google Fiber and Why It Is So Awesome." Digital Trends. http://www.digitaltrends.com/computing/what-is-google-fiber-and-why-is-it-so-awesome/#!2T8W8. Accessed June 23, 2014.
33. "The Future of Fiber." (n.d.). Google Fiber. https://fiber.google.com/newcities/. Accessed June 24, 2014.

34. Blanchard, Lauren. (2014). "Superfast Internet Makes Kansas City Neighborhood a Hotspot for Startups." *Fox News*. http://www.foxnews.com/tech/2014/06/23/superfast-internet-makes-kansas-city-neighborhood-hot-spot-for-startups/?intcmp=features. Accessed June 24, 2014.
35. Brafman, Ori and Brafman, Rom. (2010). *Click: The Magic of Instant Connections*. New York: Crown Business.
36. "Car Fires in Chaotic Vancouver after Loss." (2011). ESPN. http://sports.espn.go.com/nhl/playoffs/2011/news/story?id=6666608. Accessed June 24, 2014.
37. "Stanley Cup Riot Investigations." (2013). LEIM Conference Workshop Presentation. http://www.theiacp.org/Portals/0/pdfs/LEIM/2013Presentations/2013%20LEIM%20Conference%20Workshop%20-%20Operational%20Track%20-%20Vancouver%20Riot%20Investigation.pdf. Accessed June 24, 2014.
38. "Vancouver Riots 2011: Help Identify Suspects." (2014). Vancouver Police Department. https://riot2011.vpd.ca/. Accessed June 24, 2014.
39. Sanchez, Raf. (2013). "Boston Marathon Bombings: How Social Media Identified the Wrong Suspects." *The Telegraph*. http://www.telegraph.co.uk/news/worldnews/northamerica/usa/10006028/Boston-marathon-bombings-how-social-media-identified-wrong-suspects.html. Accessed June 25, 2014.

7

Politicization of Response and Recovery

Narrative is the beginning of recovery.

~Amanda Ripley [1]

GOVERNMENT SUPPORT OF RECOVERY

Disasters impact communities of all sizes by impacting quality of life through damage and infrastructure impacts as well as injuries and loss of life. These disasters can be caused by natural hazards like storms, flooding, and earthquakes or human-caused events such as technological failures and terrorism. Depending on the size and scope of the actual disruptive event, a local community (and the individuals and families that make it up) may be disrupted for a short period of time (e.g., days) or a longer period of time (e.g., months or years) before their lives are restored to predisaster or similar conditions. This period of recovery is a critical period of time and is filled with physical, financial, and psychological hurdles which are often complicated and unpredictable.

Because the processes of personal and community recovery are incontrovertibly linked, government ultimately plays a significant role in the recovery process. In most cases, local municipal government agencies have limited resources and personnel and therefore have difficulty restoring basic financial and economic stability—much less providing support for the individuals or businesses impacted within this community. Consequently, the consideration or obligation to help those citizens through individual assistance or those businesses through public assistance falls to the state government in the area impacted. Unfortunately, state governments can often have limited resources too—particularly during widespread geographic events. Therefore, the federal government typically carries the brunt of the financial and economic support for individuals and community businesses significantly impacted by disasters.

This capability model has been in effect for more than 200 years in the United States. For example, 19 days after a large fire burned through much of Portsmouth, New Hampshire, in 1802, Congress suspended bond payments for several months for local merchants affected by the fire. Likewise, major fires in New York City in 1835 and Chicago in 1871 triggered additional ad hoc legislation by Congress related to the financial obligations of disaster survivors (see Figure 7.1). However, major disasters like the Galveston hurricane in 1900 and the San Francisco earthquake in 1906 (which both remain two of the deadliest disasters in American history) triggered only "token aid" to both cities [2].

Figure 7.1 An artist's rendering of the Great Chicago Fire of 1871 that triggered ad hoc legislation by Congress for disaster relief. (Source: *Harper's Weekly*.)

However, this process of predominantly federal support for disaster recovery is a relatively recent development. Previous to 1950, disaster assistance in the United States was a "piecemeal approach" with federal government funding on an event-by-event basis with scope and political relevance to a particular event playing a significant factor. This process began to change with the passage of the Federal Disaster Relief Program in 1950, which transferred power from Congress to the president of the United States to federally declared disasters and more clearly established how the federal government would supplement local and state recovery efforts [3]. This process was further modified in the 1960s with the establishment of the Federal Disaster Assistance Administration within the US Department of Housing and Urban Development. This particular structure was heavily tested in response to the Anchorage, Alaska, earthquake in 1964 [3] (Figure 7.2).

Figure 7.2 In response to the Anchorage, Alaska, earthquake in 1964 the Federal Disaster Assistance Administration was formed within the US Department of Housing and Urban Development (HUD). (Source: US Army.)

Unfortunately, by 1969, the impact of Hurricane Camille led to major shifts in government support of disaster relief. Specifically, minorities and lower socioeconomic classes claimed that voluntary and humanitarian agencies active in disaster support were not providing equal support in disaster resources [3]. By 1974, Congress passed the Disaster Relief Act, which established the Federal Disaster Assistance Program which outlined procedures for requesting federal funds [4]. While the formality of government response was increasing, there were still issues—particularly around hazardous materials events and the preparedness and protection of nuclear power plants. Specifically, by the end of the decade more than 100 different federal departments shared responsibility for various components of response and recovery. Consequently, the National Governor's Association (NGA) urged US President Jimmy Carter to centralize these emergency management responsibilities (see Figure 7.3) [2]. To meet the growing needs and the political requests, President Carter established

Figure 7.3 President Jimmy Carter centralized federal emergency management responsibilities into what became the Federal Emergency Management Agency (FEMA). (Source: Marion S. Trikosko.)

the Federal Emergency Management Agency (FEMA) through Executive Order 12127 and later Executive Order 12148, which shifted various disaster relief efforts to the newly created agency [5].

However, the role of government (particularly state and federal entities) was fully established with the passage of the Stafford Act by Congress in 1988. This piece of legislation authorized funding to implement hazard mitigation measures through a national Hazard Mitigation Grant Program, which authorized the US federal government to contribute up to 75% of costs associated with hazard mitigation and restoration measures initiated in disaster-impacted areas [2]. More importantly, the Stafford Act also established the rules and guidelines for presidential disaster declarations (PDDs) which, when established, released comprehensive federal assets and resources as well as additional funding for local and state entities. Typically, this activation occurs after the results of a major disaster can be assessed; however, in some events with significant forecasted warning (e.g., Hurricane Katrina and Superstorm Sandy), the PDD can be initiated prior to the actual impact of the event [6]. There are a significant number of roles and stipulations for the activation of these programs, but almost any rule or stipulation can be altered by presidential review and approval [7].

In Other Words...Need for Transparency in Disaster Declaration Process

The move towards greater transparency may have shifted the way in which FEMA makes recommendations to the President as to whether incidents are worthy of federal assistance. Prior to the move for greater transparency, FEMA officials have private discussions to evaluate a range of factors when determining a state's financial capacity to respond to an incident without federal assistance. These factors could include the state's economic well-being, [and] whether the state had a budget surplus, among others. These factors are often subjective and difficult to quantify, which in turn make the rationale for certain recommendations more difficult to justify.

~Congressional Research Service [8]

Since the initial establishment of formal government support in the early 1950s, major disaster declarations have averaged approximately

35 per year. However, as the formality of federal government support of disaster recovery has increased, so has the number of federal disaster declarations—particularly over the last two decades. For example, from 1990 to 1999 there were an average of 46 major disasters per year; however, from 2000 to 2009 the number of disasters increased to 64 per year [8]. The highest years were 1996 and 2008, which each had 75 declared disasters [9]. Moreover, there has been a slight increase in the number of PDDs issued in presidential election years. Specifically, scholars have noted that since 1972 the sitting US president has been more likely to approve a disaster declaration in an election year than in a nonelection year. Statistics indicate a slight increase in disaster declarations and a minimal decrease in the number of declarations rejected during those presidential election cycles. Specifically, the numbers of approved and denied declarations during a nonelection year were 46.64 and 11.93, respectively. In contrast, during election years those numbers jumped to 61.8 and 7.6, respectively [8]. At the same time, other critics have argued that the number of election cycle years is too few to draw definitive connections.

This trend of increasing disasters and related declarations cannot definitively be assigned to a cause. Scholars and emergency management practitioners have leveraged these data to consider a variety of issues including possible global climate changes (see Chapter 12). However, this chapter will primarily focus on the likelihood that these trends are connected to political influence—particularly in light of growing partisanship in politics within the United States and various parts of the world. Other studies have shown that "the best predictor of a presidential disaster declaration, bar none, is actual need" [10]. However, the question—and potential opportunity for external influences—is how marginal disasters are handled.

CURRENT POLITICAL IMPLICATIONS

To understand the potential political impacts on disaster management, a brief overview of current political trends and behaviors is appropriate. In 2004, approximately 10% of Americans identified themselves as uniformly liberal or conservative. By 2014, those identified as "ideologically consistent" doubled to 21% with similar or shared views on size and scope of government, environment, and foreign policy [11]. Likewise, in 2012, the Pew Research Center conducted a 25-year study that found that the partisan gap in opinions on over 40 political values had doubled since the last quarter-century [11].

Interestingly, this partisan shift is not symmetrical between conservative (Republican) and liberal (Democrat) perspectives. Specifically, the number of Democrats who are mostly liberal has nearly doubled from 30% in 1994 to 56% in 2014 and the number of overwhelmingly liberal democrats has quadrupled from 5% to 23% over that same period of time. The shift in Republicans has been less severe, with 45% deemed as moderately Republican and another 13% as consistently conservative in 1994. However, the shift by 2014 was only up to 53% for moderate and 20% for consistently conservative [11].

This change is certainly not limited to political views of the general public. Elected officials from both parties have moved farther away from moderate. For example, the negative view of the opposing party has shifted significantly over the past 20 years. For example, in 1994, 68% of Republicans and Republican leaners had an unfavorable opinion of the Democratic Party and only 17% had an extremely unfavorable opinion. Similarly, in 1994, 57% of Democrats and Democratic leaners viewed Republicans unfavorably and only 16% had a very unfavorable view. However, by 2014 the "deeply negative" views have risen to 38% of Democrats and 43% of Republicans [12]. In both political parties, most of those who viewed the other party's policies as unfavorable felt that they "are so misguided that they threatened the nation's well-being" [12].

As the public political opinion shifts, so does the approach and perspective of democratically elected officials. While control of the legislative and executive branches has shifted back and forth between the two major American political parties over the last 20 years, the growing divide has undermined the natural compromise capability between two opposing views, has which significantly impacted the ability to pass and approve legislation. Specifically, one observer estimated that current legislative gridlock impacts approximately 75% of the salient issues under consideration [13]. This gridlock has been even more often present on issues with a smaller compromise range—or, more specifically, those issues that tend to be "black or white" ideologically are often extremely difficult for the growing partisanship to find compromise [13].

Interestingly, this legislative gridlock has lessened during some major emergencies or disasters like the terrorist attacks on September 11, 2001. For example, the 107th US Congress, which served in 2001 and 2002 under both Republican and Democratic leadership, left only 34% of political issues unresolved through legislation. Moreover, a significant number of these political issues were ultimately directly related to the disaster itself. For example, the passage of the PATRIOT Act and the authorization to use military force against terrorism maintained political cooperation at levels unmatched by other periods of recent history (see Figure 7.4) [14]. However, clearly this level of cooperation is extremely uncommon in modern times and erodes quickly after disasters.

These significant differences in political ideology are not just limited to the general public and within parties at the federal level. Specifically, there are significant political disagreements at the state level between governors and state legislative branches as well as between governors and the federal government, regardless of what party is currently holding the office. The pressure on governors is particularly evident during disaster recovery and response as "disaster assistance is an almost perfect political currency. It serves humanitarian purposes that only the cynical academic could question" [15]. More specifically, as discussed earlier, the disaster assistance funds often come from supplemental appropriations and therefore often do not count against budgetary issues (e.g., deficits) and ultimately promote the locally impacted economies. Likewise, the level of funding provided from the federal government and passed to locally

Figure 7.4 President George W. Bush signed the initial PATRIOT Act, which authorized the use of military force against terrorism. (Source: US Department of Defense/R. D. Ward.)

impacted communities through the representative state does not always match the actual scope and scale of the inflicted disaster. This unbalanced approach can be caused for a variety of reasons, but more often than not results from "the heat of media coverage of incidents," which adds to the "importance of exhibiting political responsiveness" [15]. Consequently, governors are pivotal in controlling and/or influencing the level of political influence on disaster recovery and assistance.

In Other Words...Impact of Governors on Disaster Recovery Assistance

While media and political pressure may have some influence on the outcome of some requests, governors may exercise caution since they are reluctant to be turned down when requesting aid. A denial of their request could be perceived by some to reflect adversely on their decision-making skills and judgment under pressure. Unlike the procedures of the federal process, a governor's decision to request a declaration is a public and often newsworthy action.

~Congressional Research Service [15]

This political tension felt by governors leading states impacted by emergencies or disasters has gone in many different directions over the years, but under the guidance of President Barack Obama, Republican governors have received the most attention for their political interaction during major disasters. For example, in 2010, Louisiana Governor Bobby Jindal portrayed President Obama as disconnected from the Gulf of Mexico oil spill response and recovery efforts due to being "more focused on the political aftermath than the actual impact of the crisis" [16]. Specifically, Jindal described how the president complained about a letter Jindal had sent to the administration requesting authorization for food stamps for those who had lost employment due to the oil spill due to concern about how it might look politically [16] (see Figure 7.5). The politicization of this particular disaster continued into 2012 when Obama and Jindal publicly argued over the use of foreign companies who were offering to continue to assist in the cleanup in the Gulf. Specifically, Jindal objected to Obama's refusal to waive the ban on allowing international support in American waters [17].

On the other hand, some Republican governors have taken a positive political connection with President Obama in response to governmental

Figure 7.5 Louisiana Governor Bobby Jindal complained that President Obama was too concerned with the political implications from the 2010 BP oil spill. (Source: White House/Pete Souza.)

disaster aid and recovery. For example, New Jersey Governor Chris Christie had a defining political moment in how he interacted with the Obama administration's response to Superstorm Sandy in 2012. Specifically, Christie told local and national media agencies that President Obama's response had been "outstanding" and "wonderful" and that the president "deserve[d] great credit" and did a "great job for New Jersey" (see Figure 7.6) [18]. Interestingly, Christie's response to the storm was well received by his constituency as his approval rating jumped from 56% before the storm to 72%–77% after (depending on the poll) [19]. This level of bipartisan politics is certainly not common, as was seen by Governor Jindal's response during the BP oil spill. In both cases, pundits have argued that the gubernatorial actions toward Obama were more related to personal political motivations (e.g., possible future presidential runs); however, regardless of the reason it continues to show a strong trend toward increasing politicization during disaster recovery even at the state level.

The impact of political choices can also have cascading impacts on governors in other areas. For example, as described earlier, the traditional approach to governmental disaster assistance during PDDs is for the federal government to provide reimbursement for 75% of related and approved costs while the impacted state and local government agencies split the remaining 25% of recovery costs. However, there have been major

Figure 7.6 New Jersey Governor Chris Christie told national media agencies that President Obama's response to Superstorm Sandy was outstanding and he ultimately shared political credit for the disaster recovery. (Source: White House/ Pete Souza.)

disasters like the massive tornado outbreak that struck Alabama in 2011 that triggered altered approaches to federal assistance. Specifically, the federal government initiated a pilot program called Operation Clean Sweep that authorized a 90% federal cost share for the removal of debris in counties designated for federal assistance that had extensive damage [20]. While this program was ostensibly implemented to improve the pace and effectiveness of debris removal, it brought inherent political connectedness not only during the recovery in Alabama, but also for major events that occurred afterward. In essence, pilot programs like Operation Clean Sweep or alterations to the federal assistance model risk becoming precedent and thus creating even more areas of political strife and economic burden during future events.

Similarly, the significant impacts of Superstorm Sandy on portions of the northeast United States also triggered exceptional changes to federal aid programming. Specifically, to help individuals cover expenses incurred in the immediate aftermath of the storm (e.g., rent), FEMA allocated more than $1.4 billion in assistance to more than 182,000 individuals and homeowners in the five states where the storm damaged or destroyed more than 650,000 structures. These funds were dispersed fairly quickly, but grants for an additional $1.7 billion

($150,000 to $300,000 per household) were provided through the US Department of Housing and Urban Development (HUD) to help rebuild storm-damaged homes. Unfortunately, the distribution of these funds was delayed by more than 1 year. HUD representatives stated that funds were available, but only after applicants (also known as disaster survivors) met a multitude of federal requirements implemented after Hurricane Katrina to mitigate against the duplication of monetary benefits and overall abuse of the disaster assistance [21]. While most likely wholly caused by government bureaucracy, delays such as these in the delivery of disaster aid can often cause significant friction for state and local governments that must deal with the survivors more directly.

In Other Words...Federally Mandated "Hoops" for Disaster Assistance

HUD requires eligible homeowners to jump through numerous federally mandated hoops—much of which we and the governor[s] have felt is unnecessary and redundant given that these are existing homes that cannot be built bigger than what was there before...Each of these steps take time and may get complicated based on each homeowner's individual circumstance.

~New Jersey Department of Consumer Affairs [21]

FUTURE POLITICAL ISSUES

As comprehensively established earlier in this chapter, government (particularly federal government) has a unique and critical role in the provision of disaster relief aid. Under traditional roles, there is a very sensitive political balance between local, state, and federal components as well as in the current party establishment which is increasingly partisan and divided. Given this structure, recovery activities are often highly political with as much as 50% of all disaster relief being politically motivated or influenced [22]. However, the recovery process is not the limits to which emergency management and disaster readiness activities are influenced by political decisions and partisan leanings and perceptions—specifically, current "hot button" issues, like energy generation, climate change, immigration, gun control, drones, and the socioeconomic divide between the so-called "rich" and "poor." The remaining portion of this chapter will look at each of these issues in greater detail.

The first issue under consideration will be the generation of energy in a world aware of and acting upon global warming and other related climate issues. In 1973, a scientist in the *Journal of Geophysical Research* first asked the question about whether the global climate was changing in such a way to contribute to secondary effects like the disintegration of the Antarctic ice sheet [23]. However, it was not until 1979 that the first World Climate Conference was held. That conference represented the first significant global review by the United Nations, which ultimately determined that "carbon dioxide plays a fundamental role in determining the temperature of the earthquake's atmosphere and it appears plausible that an increased amount of carbon dioxide in the atmosphere can contribute to a gradual warming...but the details of the changes are still poorly understood" [24] (see Figure 7.7). By 1988, the United Nations established the Intergovernmental Panel on Climate Change (IPCC) to assess "available scientific data and the possible broader impacts of climate change."

Although slow in developing, these concerted global efforts led to the establishment of the Kyoto Protocol in 1997 that required 37 industrialized nations (including the United States) to reduce greenhouse gas emissions due to the fact that these nations shared more responsibility for the current level of pollution than developing countries [24]. Unfortunately, many

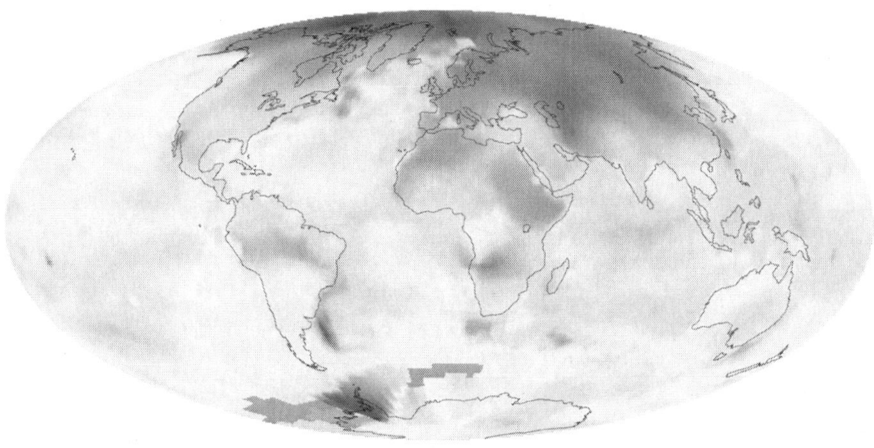

Figure 7.7 Governmental reviews have begun to embrace the concept that global temperatures are rising due to human behavior. (Source: NASA/Robert Simmon.)

critics of the Kyoto Protocol argued that the science justifying the claims was unreliable and that it ultimately overstated its claims. Moreover, the Russian Academy of Sciences publicly stated that the Russian government's signing of the protocol was purely a political move rather than globally altruistic. US President George W. Bush was also an avid critic, repeatedly refusing to sign the treaty. Bush's view was ultimately consistent with that of the US Senate, which voted unanimously against signing any treaty that would cause "severe economic damage to the United States, while exempting the rest of the world." Specifically, the Bush administration claimed the Kyoto Protocol would cost the American economy $400 billion and 4.9 million jobs [25].

In Other Words...The Impact of the Kyoto Protocol

Overall, Kyoto appears to be a restrictive and ineffective agreement. The battle over whether the United States should sign or not appears to be more political than practical. While most agree something should be done to ensure the health of our planet, entering into a flawed agreement is not the answer. Signing the Kyoto Protocol could stand in the way of finding better, more effective alternatives.

~Jim Smoot [25]

However, the political objections to global warming and climate change did not continue. After losing his bid for the US presidency in 2000, former senator and Vice President Al Gore launched a campaign and related video documentary (*An Inconvenient Truth*) to provide public education about the dangers of global warming. Upon receiving positive critical reviews and ultimately two Academy Awards, Gore became the "face" of the need for climate change [24]. Not surprisingly, given the high-profile nature of Gore's efforts to bring attention to the importance of climate change, the opinion of Americans believing there was "solid evidence" that human activity was the primary cause of Earth's warming temperatures rose to 41% [24].

By 2007, major energy providers, like ExxonMobil, announced they were discontinuing their financial support of research groups that were questioning climate change. However, nearly simultaneously, other major donors like the American Petroleum Institute and various other private donors began making significant contributions to these same groups.

For example, a nonprofit research organization called the Heartland Institute held its first annual conference for skeptics of climate change [24]. Even though some bipartisan support from well-known politicians like Newt Gingrich, Nancy Pelosi, and John McCain was announced in support of climate change by 2008, the concept of supporting or denying climate change became highly politically contentious with liberal Democrats supporting the need for action to address the implications of climate change, while conservative Republicans nearly uniformly held the opposing view of this particular issue (see Figure 7.8).

The political tension over this issue impacted nongovernmental bodies as well. For example, an open letter from the National Academy of Sciences released in 2010 called "for an end to McCarthy-like threats of criminal prosecution against our colleagues based on innuendo and guilt by association, the harassment of scientists by politicians seeking distractions to avoid taking action and the outright lies being

Figure 7.8 Even though significant conservative leaders like John McCain announced support of climate change, Republicans remain nearly uniformly opposed to this issue. (Source: US Congress.)

spread about them" [24]. Their announcement corresponded with the US midterm congressional elections, which resulted in a shift toward Republican control with a large "Tea Party" contingent, which, generally speaking, staunchly opposed climate change issues [24]. Moreover, with the convening of that Congress, Republican leadership also eliminated the House Select Committee on Energy Independence and Global Warming, which had been established in 2007 by Democratic House Speaker Nancy Pelosi. Republicans nixed the committee as part of a philosophical political pledge to "get rid of waste and duplication in terms of how we run the Congress...we'll save several million dollars" with the intention of allowing the standing Science Committee to address the issue [26]. This stance was further countered in 2012 by the Democratic Party, which altered its platform to acknowledge climate change as "one of the biggest threats of this generation" and by President Obama, who stated that "climate change is not a hoax.... [It is] a threat to our children's future" [24]. The politicization of climate change shows no indications of slowing down and will continue to influence decision makers and politicians, which will in turn impact how emergency managers address climate change and related risks, which will be further evaluated in Chapter 12.

As an integral issue to climate change, the politicization of how energy is used and generated is critical to understand. Politicians on both side of the issue have strong and public opinions about traditional energy generators such as coal, petroleum, and nuclear as well as so-called green energies such as wind, solar, and other hybrids. These opinions are based on environmental impacts and economic concerns, as well as perceived and real risks to local communities (and in some cases, the world) from these energies that might cause an emergency or disaster. Typically, conservative politicians prioritize the importance of these energy generators as sources of jobs and economic stimulation, while more liberal politicians stress the risk (both environmental and catastrophic) of traditional energy sources and strongly promote green energy.

The use of petroleum-based sources of energy includes crude oil, coal, and natural gas, which are formed from prehistoric matter of plants, animals, and other life forms that were buried and ultimately exposed to high pressure over millions of years. These so-called fossil fuels represent the major source of energy throughout the world—including as much as 80% of energy consumption in the United States [27]. Interestingly, the United States produces and consumes the most crude oil in the world. Specifically, in 2009, the United States imported 51% of the oil it used and

Figure 7.9 Oil drilling and commercial production take place in many different areas of the United States. (Source: Eric Kounce.)

consumed 22% of the world's petroleum. Oil is produced in 31 states and offshore in many parts of the United States (see Figure 7.9). Likewise, coal is the most abundant fossil fuel in the United States and generates nearly half of all electricity domestically and more than 40% of the global electricity [27]. Unfortunately, energy production—particularly in the United States—is a contentious issue, as drilling, mining, and related transportation activities have contributed to substantial air pollution and ground and surface water pollution.

Petroleum-based energy products can also be dangerous to remove from the ground and are often located in environmentally sensitive areas such as remote forested areas or in major bodies of water. For example, in the late spring of 2010, an oil drilling platform operated by BP in the Gulf of Mexico called the Deepwater Horizon exploded due to an oil leak. Emergency responders and BP officials quickly determined that the leak was not isolated, but rather was leaking from the drill shaft deep under the water's surface. After a prolonged response and related recovery over much of that summer, approximately 206 million gallons of oil leaked from the well, which represented more than 19 times the amount of oil that leaked from the wrecked Exxon *Valdez* in 1989 and was worth nearly $400 million at current market prices [28].

In Other Words...Political Fallout from BP Oil Spill

In 2008, a bipartisan agreement was reached to lift the decades-long ban on new offshore [oil] drilling and open new areas off the Atlantic, Pacific and Arctic coasts. Since President Obama took office he has systematically taken steps to re-impose an offshore drilling moratorium.... No new drilling...will occur during [his] term in office—despite the overwhelming support of the American people for new offshore energy production.

~Doc Hastings [29]

In response to the Deepwater Horizon oil spill, President Barack Obama received significant political pressure from constituents, media, and both political parties. To defer this pressure, his administration stressed BP's responsibility, stating that "we will hold them fully accountable on behalf of the United States" and implemented various executive policies to ostensibly reduce risk from this type of energy production [30]. Specifically, a moratorium on deepwater drilling permits in US waters was extended for 6 months and land leases for drilling were suspended in Alaska, Virginia, and parts of the Gulf of Mexico. However, as discussed earlier, conservatives and Republicans argued these moratoriums were influenced by political priorities rather than economic or widespread safety concerns.

A similar political battle occurred in 2011 after a 9.0 moment magnitude (M_W) earthquake and related tsunami caused the meltdown of the Tokyo Electric Power Company's (TEPCO) Fukushima Daiichi nuclear power plant. Specifically, the flood water overwhelmed the reactor generators, which caused them to fail and raise the temperature of the reactor cores and ultimately melt down. An evacuation order was given early in the process, but the long-term impact to the surrounding areas was not avoided [31]. In the now deserted town where the plant was located, the ambient radiation is 90 times higher than is considered safe, 85,000 people are still under mandatory evacuation order from the areas of highest radiation, and more than 390,000 homes have been identified as needing decontamination [32]. In response, Japanese authorities initially shut down all 54 national nuclear power plants, which was publicly popular as antinuclear sentiments run deep in the area. However, with growing power shortages and a shift to a pronuclear government leadership, Japan began to reopen many of these facilities by 2013 [33].

Given that radiation from the Fukushima nuclear power plant slowly moved across the Pacific and reached portions of Canada and the northwestern United States by 2014, political pressure on nuclear power generation has increased in North America as well [34]. Unlike Japan, the United States has utilized strong regulatory limitations and political pressure to significantly reduce the growth of nuclear power generation. For example, no new nuclear power plants have been generated in nearly 30 years with estimates that proposed plants in Georgia and South Carolina will take at least 6 more years to complete [35]. However, after Fukushima and other historical incidents like the meltdowns at Three Mile Island and Chernobyl, many communities (and their respective political leaders) are very concerned about the presence of these nuclear facilities in their communities. It is under this sensitive political and public balance that emergency managers must continue to address real and perceived risk from nuclear energy production, including appropriate buffer zones and long-term storage processes.

In Other Words...Perception of Risk from Nuclear Energy Production

"Fukushima" now conjures up what "Chernobyl" did for a previous generation: scenes of catastrophic accidents that turn everyday power generation into serious health hazards and costly cleanups that last for decades. The 2011 meltdown at the Fukushima Daiichi plant carried enormous political impact as well. Japan shut down all its nuclear plants, Germany swore off nuclear energy and the United States rushed to review safety at all its reactors.

Yes, other big industrial operations also have major and sometimes fatal accidents; some, such as oil spills or chemical explosions, have effects lasting years. But because nuclear reactors use radioactive materials, the potential danger of nuclear accidents is greater—as is the public perception of risk.

~Keith Johnson [35]

Active-shooter events represent another politically sensitive disaster response and recovery effort. According to the US Federal Bureau of Investigation (FBI), there was a steady increase in the number of active-shooter events from 2000 to 2012 with a particular increase in the frequency after 2008. Specifically, the frequency went from one every other month prior to 2009 to one per month from 2009 to 2012. Moreover, the number of people shot per event has also steadily risen over that same period of time [36]. The location of these active-shooter events is also diverse, with incidents at businesses, schools, outdoors, and/or a combination of sites. Despite the significant media attention and related public concern, active-shooter events at schools only occurred 29% of the time, which was second to business locations [36].

After nearly every active-shooter event (including Columbine High School, Gabby Gifford, Aurora theater, and Newtown), political divisiveness over gun control gets significant attention. Democrats and progressives tend to try to leverage these catastrophes to call for restrictions on gun sales with some limitations related to capabilities, ammo sizes, or types of background checks. These political discussions are often played out in the media and via online forums more so than actual changes. In reality, public support for increased gun control often peaks immediately following an active-shooter event and then wanes precipitously as time and attention move elsewhere. While some states, like New York, Connecticut, and Colorado, have passed statewide laws limiting certain aspects of gun ownership, federal-level control has rarely gained any traction.

Interestingly, the other major component that has increasingly received attention after active-shooter events is the presence of mental health issues for many of the shooters. The FBI analysis of active-shooter events has routinely described the perpetrators of active-shooter events as "social isolates" who "harbored feelings of hate and anger" and often had some contact with mental health professionals. However, few of these individuals had previous arrests for violent behavior. Rather, they typically experienced significant emotional hardship such as changes in personal and/or intimate relationships, financial status, employment, or housing [36]. Most experts agree that emergency managers and public safety officials must find an appropriate intersection between mental health awareness, physical security, and training of the public to create an appropriate synergy toward the reduction of risk from active-shooter events [37]. However, much like the push for gun control, there is significant political hesitation to commit the resources and political pressure needed to implement an appropriate level of mental health programming to mitigate active-shooter events. For example, President Barack Obama's proposed $150 million "mental health first aid" legislation for schools stalled in Congress. Likewise, US federal funding for the National Institute of Mental Health was reduced by more than $12 million, which was called "an extremely disturbing development, given the overall public-health burden of [various mental health] disorders" [38]. Clearly, extra funds for emergency management mitigation efforts cannot be fully implemented when baseline political funding is already lacking.

These major emergency management issues only begin to scratch the surface on how politics is leveraged during disasters for political favor, gain, or discourse. Rising issues such as homeland security risks from border protection and immigration as well as the use of drones for militaristic, public safety, and commercial services will continue to be widely discussed in communities and therefore by the politicians that represent them. If the political structure of disaster recovery as well as the historical examples of political response is any indication, emergency managers cannot avoid the impact of politics during disaster response and recovery. Consequently, professional responders of all types must embrace the possibilities—both positive and negative—of how the emergency management process may continue to be shaped in the future by this process.

REFERENCES

1. Ripley, Amanda. (2009). *The Unthinkable: Who Survives When Disaster Strikes and Why*. New York: Harmony.
2. "The Federal Emergency Management Agency." (2010). Federal Emergency Management Agency (FEMA). https://www.fema.gov/pdf/about/pub1.pdf. Accessed June 27, 2014.
3. "History of Disaster Relief." (2012). *Texas Impact*. http://www.texasimpact.org/Disaster_Relief_History. Accessed June 27, 2014.
4. "Federal Disaster Assistance Program." (1998). University of Florida Cooperative Extension Service. Chapter 3. http://disaster.ifas.ufl.edu/PDFS/CHAP03/D03-09.PDF. Accessed June 27, 2014.
5. "FEMA: 35 Years of Commitment." (2014). Federal Emergency Management Agency (FEMA). https://www.fema.gov/fema-35-years-commitment. Accessed June 27, 2014.
6. "Hurricane Sandy: Schumer Calls for Pre-emptive Disaster Declaration." (2012). *The Yeshiva World News*. http://www.theyeshivaworld.com/news/headlines-breaking-stories/142852/hurricane-sandy-schumer-calls-for-preemptive-federal-emergency-decleration.html. Accessed June 28, 2014.
7. "Understanding the Stafford Act: Its Effect on Public Entities." (n.d). *PrimaCentral*. http://www.primacentral.org/resources/Stafford%20Act%20FAQs.pdf. Accessed June 28, 2014.
8. Lindsay, Bruce R. and McCarthy, Francis X. (2012). "Stafford Act Declarations 1953–2011: Trends and Analyses and Implications for Congress." Congressional Research Service. http://fas.org/sgp/crs/homesec/R42702.pdf. Accessed June 28, 2012.

9. Kukich, Diane. (n.d.). "Presidential Disaster Declarations on the Rise, National Expert Says." University of Delaware Research. http://www.udel.edu/researchmagazine/issue/vol2_no1_enviro/disasterdeclarations.html. Accessed June 28, 2014.
10. Bluth, Gregory. (n.d). "Politics and Professional Culture." Michigan Tech University. http://www.geo.mtu.edu/~gbluth/Teaching/GE4150/lecture_pdfs/L5_politics.pdf. Accessed June 29, 2014.
11. Shedlock, Mike. (2014). "Time Lapse Images of Growing US Political Politicization; Root Cause of the Shrinking Middle Class." *MISH's Global Economic Trends*. http://globaleconomicanalysis.blogspot.com/2014/06/time-lapse-image-of-growing-us.html. Accessed June 29, 2014.
12. "Section 2: Growing Partisan Antipathy." (2014). Pew Research Center for the People and the Press. http://www.people-press.org/2014/06/12/section-2-growing-partisan-antipathy/. Accessed June 30, 2014.
13. Binder, Sarah A. (2014). "Polarized We Govern?" Brookings Institution. http://www.brookings.edu/research/papers/2014/05/27-polarized-we-govern-congress-legislative-gridlock-polarized-binder. Accessed June 30, 2014.
14. Binder, Sarah. (2014). "Polarized We Govern?" Center for Effective Public Management at Brookings. http://www.brookings.edu/~/media/research/files/papers/2014/05/27%20polarized%20we%20govern%20binder/brookingscepm_polarized_figreplacedtextrevtablerev.pdf. Accessed June 30, 2014.
15. McCarthy, Francis X. (2011). "FEMA's Disaster Declaration Process: A Primer." Congressional Research Service. http://fas.org/sgp/crs/homesec/RL34146.pdf. Accessed July 1, 2014.
16. Martin, Jonathan. (2010). "Bobby Jindal Hammers Barack Obama in New Book." *Politico*. http://www.politico.com/news/stories/1110/45021.html. Accessed July 1, 2014.
17. Picket, Kerry. (2012). "Picket: Flashback—Obama Quested over Week Louisiana Response during 2010 BP Oil Spill." *Washington Times*. http://www.washingtontimes.com/blog/watercooler/2012/sep/1/picket-flashback-obama-questioned-over-weak-louisi/. Accessed July 1, 2014.
18. McGregor, Jena. (2012). "In Superstorm Sandy, Governor Christie Praises Obama's Crisis Leadership." *Washington Post*. http://www.washingtonpost.com/national/on-leadership/in-superstorm-sandy-new-jersey-governor-chris-christie-praises-president-obamas-crisis-leadership/2012/10/30/89769e32-22b5-11e2-ac85-e669876c6a24_story.html. Accessed July 1, 2014.
19. Sullivan, Sean. (2013). "How Superstorm Sandy Became Christie's Defining Moment." *Washington Post*. http://www.washingtonpost.com/blogs/the-fix/wp/2013/10/29/how-superstorm-sandy-became-chris-christies-defining-moment/. Accessed July 1, 2014.
20. "Gov. Bentley Asks for 60 Day Extension of Operation Clean Sweep." (2011). WAFF 48. http://www.waff.com/story/15015035/gov-bentley-asks-for-extension. Accessed July 2, 2014.

21. Leitsinger, Miranda. (2013). "Where Is All the Money?: Pace of Aid Efforts Frustrates Sandy Survivors." *NBC News*. http://www.nbcnews.com/news/other/where-all-money-pace-aid-effort-frustrates-sandy-survivors-f8C11487331. Accessed July 2, 2014.
22. Wilson, S. et al. (2013). "The Lack of Disaster Preparedness by the Public and Its Affect on Communities." *Internet Scientific Publication*. http://ispub.com/IJRDM/7/2/11721. Accessed July 2, 2014.
23. "Collapse or Catastrophe?" (2014). *The Economist*. http://www.economist.com/news/science-and-technology/21602190-west-antarctic-ice-sheet-looks-doomedeventually-collapse-or-catastrophe?zid=313&ah=fe2aac0b11adef572d67aed9273b6e55. Accessed July 3, 2014.
24. Childress, Sarah. (2012). "Timeline: The Politics of Climate Change." PBS. http://www.pbs.org/wgbh/pages/frontline/environment/climate-of-doubt/timeline-the-politics-of-climate-change/. Accessed July 3, 2014.
25. Smoot, Jim. (2007). "The Case against the Kyoto Protocol." *World Issues 360*. http://www.worldissues360.com/index.php/the-case-against-the-kyoto-protocol-2-76190/. Accessed July 3, 2014.
26. Sonmez, Felicia. (2010). "House Republicans Nix Global Warming Committee." *Washington Post*. http://voices.washingtonpost.com/44/2010/12/house-republicans-nix-global-w.html. Accessed July 6, 2014.
27. "Petroleum, Natural Gas, and Coal." (2012). *Burn and Energy Journal*. http://burnanenergyjournal.com/petroleum-natural-gas-and-coal/. Accessed July 6, 2014.
28. Repanich, Jeremy. (2010). "The Deepwater Horizon Oil Spill by the Numbers." *Popular Mechanics*. http://www.popularmechanics.com/science/energy/coal-oil-gas/bp-oil-spill-statistics. Accessed July 7, 2014.
29. "Obama Administration Imposes Five Year Ban on Majority of Offshore Drilling." (2011). House Committee on Natural Resources. http://naturalresources.house.gov/news/documentsingle.aspx?DocumentID=267985. Accessed July 7, 2014.
30. "Obama Defends Response to Gulf Oil Spill, Pledges to 'Shut This Down.'" (2010). *Fox News*. http://www.foxnews.com/politics/2010/05/27/obama-federal-government-charge-oil-spill-response. Accessed July 8, 2014.
31. Crowe, Adam. (2013). *Leadership in the Open: A New Paradigm in Emergency Management*. Boca Raton, FL: CRC Press.
32. Hubris, Joe. (2013). "The TEPCO Nuclear Disaster, Two Years Later." Joe Hubris Blog. http://joehubris.com/node/88. Accessed July 8, 2014.
33. Schiffman, Richard. (2013). "Two Years On, America Still Hasn't Learned Lessons of Fukushima Nuclear Disaster." *The Guardian*. http://www.theguardian.com/commentisfree/2013/mar/12/fukushima-nuclear-accident-lessons-for-us. Accessed July 8, 2014.
34. Oskin, Becky. (2014). "Fukushima's Radioactive Ocean Arrives at West Coast." *livescience*. http://www.livescience.com/43631-fukushima-radiation-ocean-arrives-west-coast.html. Accessed July 9, 2014.

35. Johnson, Keith. (2013). "What's Holding Back Nuclear Energy." *The Wall Street Journal*. http://online.wsj.com/news/articles/SB10001424052702304906704579115510865908136. Accessed July 9, 2014.
36. "FBI Analysis Finds That Only 4% of Active Shooter Incidents since 2002 Were Perpetrated by Women." (2013). *Public Intelligence*. http://publicintelligence.net/only-4-percent-of-active-shooters-were-women/. Accessed July 10, 2014.
37. Jasper, Todd. (2013). "A Cross-Disciplinary Approach to Preventing Active Shooter Events." *Continuity Insights*. http://www.continuityinsights.com/blogs/2013/07/cross-disciplinary-approach-preventing-active-shooter-incidents. Accessed July 10, 2014.
38. Johnson, Alan and Candisky, Catherine. (2013). "Mental-Health System Overwhelmed, Underfunded." *The Columbus Dispatch*. http://www.dispatch.com/content/stories/local/2013/05/26/overwhelmed-underfunded.html. Accessed July 10, 2014.

8

Distorted Perception of Risk

To invest...$4 billion to try to prevent another $39 billion in losses... seems to me to be...a pretty smart investment to make for the country.

~Chris Christie, Governor of New Jersey [1]

UNDERSTANDING RISK

It is a critical component of professional emergency management and public safety operations to be aware of various hazards and threats within a given community. This awareness is typically established through formalized assessments that look at a combination of likelihood and vulnerability to the identified hazards. For example, a landlocked community does not have to worry about the presence of tsunamis. Likewise, communities with high sea-level elevation have minimal concern for flooding. However, these examples are highly simplistic and do not take into account the true complexity of individual communities that have various levels of economic, social, cultural, and political influences that may increase or reduce the actual risk. Likewise, the perception of those same risks can significantly alter individual and community preparedness and mitigation of those risks, which may or may not be appropriate for a given area. This chapter will focus on these perceptions and how they are changing in current and future considerations.

It is critical for this chapter and all applications of emergency management and public safety to understand the difference between hazards, threats, and risks. Unfortunately, professional emergency services providers and risk managers often use these terms interchangeably when in reality they are distinctly different. The simplest definition of a hazard is anything that can cause harm, while risk is related to the likelihood and extent of harm from the given hazard [2]. In contrast, military services, intelligence, and law enforcement communities have traditionally used the term "threat" to describe human-caused events like terrorism [3]. Since traditionally the term "hazard" was limited to naturally occurring events, these terms were mutually exclusive until homeland security got significant attention globally after the World Trade Center attacks in 2001. Since that time, the terms "hazard" and "threat" have become more synonymous; however, the original delineation is still held depending on the emergency response or disaster management discipline considering the issue and assessing the risk created by these hazards and threats (see Figure 8.1).

The most traditional approach to understanding risks is to create a matrix of probability and consequence of a given hazard or threat. Typically, probability and consequence for each hazard are measured on a numbered scale (1 to 5) and cross-referenced against each other. These cross-referenced evaluations are then classified into subjective categories

Figure 8.1 Hazard classifications were mostly limited to naturally occurring events prior to the terrorist attacks of September 11, 2001. (Source: Dana Trytten.)

such as high, medium, and low. For example, a given hazard rated a 5 for both probability and consequence (as might occur for flooding in a coastal community) would be deemed extremely high. Conversely, a given hazard rated a 1 for both probability and consequence (as might occur for a tornado for an island community) would be deemed extremely low. This process is followed for all subjectively identified hazards, which in turn allows for emergency response personnel and risk managers to categorize and rank the threats and hazards that may impact a community to ensure planning and mitigation strategies are efficient and effective.

The most challenging component of the traditional risk assessment process is how to evaluate hazards with mixed ratings. For example, the threat of an active-shooter event has had catastrophic consequences in numerous communities, but is still very unlikely to occur in any given community or building. Given the traditional model of assessment, the probability and consequence of an active-shooter event would only be listed as a low- to medium-risk event. However, significant media and public attention are focused on this event as it often involves vulnerable and unique populations such as school students, military bases, business patrons, and community leaders. The perception of this risk (and similarly classified risks) is significantly different from the subjective analysis of professional risk managers. Consequently, people have increased anxiety related to these types of events and are more willing to focus preparedness and personal risk reduction toward these threats than they are toward more conditions that are of greater actual risk (see Figure 8.2).

Figure 8.2 An individual's perception of risk greatly impacts his willingness to prepare for various hazards and threats. (Source: FEMA/Tim Pioppo.)

In her book, *The Unthinkable: Who Survives When Disaster Strikes and Why*, Amanda Ripley begins to address the impact of these perceptions of how risk is assessed. In addition to the probability and consequence components established earlier, Ripley adds dread as a component. She loosely defines dread as the combination of uncontrollability, unfamiliarity, imaginability, suffering, scale of destruction, and unfairness of the given hazards and threats [4]. Each of these considerations adds credibility to the likelihood of perception being as strong as or stronger than actual risk to a given community. The active-shooter example described earlier is certainly influenced by these characteristics of dread. For example, active-shooter events are completely uncontrollable, have significant suffering, and are wholly considered unfair. Likewise, even though they are happening (potentially on an increasing basis), they are still difficult for most people to imagine or accept in a given community.

In Other Words...Purpose of Risk Assessment

The purpose of risk analysis and risk quantification is always to provide input to an underlying decision problem which involves not just risks, but also other forms of costs and benefits...[thus] risk must... be considered always within a...context.

~Stanley Kaplan and B. John Garrick [5]

RISK PERCEPTION

Understanding the hazards, threats, and processes needed to assess potential risks is important, but only a small fraction of understanding risk in a given community. According to noted disaster sociologist Dennis Mileti, "people respond to a risk or hazard in ways consistent to their perception of that risk...[which in turn] influences behavior or action" [6]. The framing of individual or family perception is based on a variety of characteristics. For example, researchers since the 1970s have known that awareness of the hazard, knowledge of how it affects the community, and related personal vulnerability to potential threats in a given community highly impact perception of risk (see Figure 8.3). However, all of these characteristics are highly subject to error by individuals based on rumor,

Figure 8.3 Researchers since the 1970s have known that awareness of the hazard, knowledge of how it affects the community, and related personal vulnerability to potential threats in a given community greatly affect the willingness of a community to be prepared. (Source: FEMA/Michelle Collins.)

speculation, inaccurate information, and personal (or lack thereof) experience. For example, Mileti has determined in research that frequent exposure to "hazard relevant information does not automatically elicit attention and comprehension, let alone acceptance, personalization, and retention required to initiate hazard adjustments" [6].

This is very common in colloquial stories shared within the emergency management communities in regard to disaster survivors in highly susceptible areas. For example, residents in hurricane-prone communities along the coasts of the Atlantic or Gulf of Mexico who previously experienced hurricanes often doubt future hurricane risk is as significant as before because they think that it cannot happen again or simply will not be as impactful. Realistically, these hazard events are each unique and therefore have similar risk valuations regardless of the timing or personal experience. Specifically, there is no reason an impactful hurricane will not impact a coastal community repeatedly. Unfortunately, professional emergency managers do not often have a clear sense of what portions of their communities actually have this perception of risk and how many have a more realistic acceptance of risk. Consequently, it can be difficult to assess whether a given community has initiated recommended suggestions for preparedness, response, recovery, and mitigation such as evacuation and modification of residences and other buildings.

As a similar example, noted social observer and writer Malcolm Gladwell uses a similar example in his book, *David and Goliath: Underdogs, Misfit, and the Art of Battling Giants*. Specifically, Gladwell talks about the German aerial bombing of London in the early part of World War II. The British government was so concerned that the general public would be devastated by these attacks that significant preparedness was implemented including the establishment of psychiatric field hospitals for distraught community members (see Figure 8.4). In the end, the bombings did come and were impactful to the city, but not in the way originally predicted. The government found that those individuals who were not killed, who literally "crawled out of the wreckage," became emotionally stronger and ultimately exhibited less and less fear about the bombings and the related consequences [7]. This type of desensitization is common with risk and can be highly impactful.

Figure 8.4 During the German aerial bombing of London in the early part of World War II, the British government was overly concerned that the general public would be devastated by these attacks. (Source: United Kingdom archives.)

In Other Words...Motivation and Understanding in Risk Perception

People do not actually need to understand the hazard in order to be motivated enough to prepare, but they need to believe that the hazard really exists and that protection is needed.

~Michael K. Lindell and Ronald W. Perry [6]

Another consideration of risk perception is the so-called cultural theory originally developed by Mary Douglas and Aaron Wildavsky. This theory posits that various risk perceptions can be explained by four distinct cultural biases: hierarchy, egalitarianism, individualism, and fatalism [8]. This theory suggests that these perceptions help delineate different social or sociological groups within a given community. The first cultural bias is hierarchical or government functions. This is the cultural perception that government provides significant support for all aspects of community life and will continue to maintain such provisions before, during, and after hazardous events. Overconfidence in government support during disasters has increased and can lead to significant difficulties, as was seen during the response to Hurricane Katrina in New Orleans in 2005 (see Chapter 6). The second cultural bias is egalitarian or the belief that equality is necessary across all cultural and community sectors (e.g., gender and race). Likewise, individualistic or market bias can also influence perception. This is the overconfidence in a person's or community's ability to recover from an event. Even in an economically and politically diverse community a business or entire marketplace can fail after a disaster. The final bias is that of cultural fatalism or that belief that preparedness and other disaster readiness activities have no impact on overall disaster impacts [9]. Much like the example from World War II, citizens in developed countries often feel that the preparedness is not worth the effort or the time (often for a variety of reasons). While these categories are generalized, they often represent major components of a culture or community and exist in tension to one another.

The final major consideration related to risk perception is the personality types of individual community members and potential disaster survivors. This is a challenging issue as sociologists and psychologists do not consistently agree on how personality traits impact disaster preparedness choices and individual safety actions. However, there seems to be a growing acceptance of the indirect effects of personality as these traits certainly can affect "perception and appraisal of the environment" [10]. This process has also been supported in social cognition models which evaluate how people acquire and maintain certain behavioral patterns [11]. Understanding how these behaviors change may be a way for professional risk and emergency managers to grasp how personality and individual choices may be impacted before, during, and after disasters.

HISTORY OF PROFESSIONAL RISK ASSESSMENT

Much like the general public, professional emergency managers and related disciplines perceive risk and prioritize resources differently too. As discussed in previous chapters, the concept of professional, government-based emergency management is a relatively new field that firmly has its roots in Cold War era civil defense. With the end of World War II and the rise of the United States and the Soviet Union as the world's two major superpowers, tensions quickly rose between these countries and led to nearly three decades of nuclear tension (see Figure 8.5).

Figure 8.5 With the rise of the United States and Soviet Union as world superpowers, tensions quickly rose between the two and led to three decades of global tension over nuclear weapons. (Source: National Nuclear Security Administration.)

Air raid sirens, "duck and cover" drills at schools, and nuclear fallout shelters in schools, residences, and government facilities became social norms and were overwhelmingly the focus of community risk managers during that time. As such, resources and planning efforts primarily focused on preparing for the nuclear risk and thus left communities more susceptible to (more likely) risks in their areas.

In Other Words...The First Nationwide Civil Defense Drill

On June 14, 1954...over 12 million Americans "die" in a mock nuclear attack as the United States goes through its first nationwide civil defense drill. The...drill was organized and evaluated by the Civil Defense Administration and included operations in 54 cities in the United States.... The basic premise of the drill was that the United States was under massive nuclear assault from both aircraft and submarines.... Each citizen was supposed to know where the closest fallout shelter was located; these included the basement of government buildings and schools, underground subway tunnels, and private shelters.... Government officials pronounced themselves very pleased with the drill.... A more cautious assessment came from a retired military officer who observed that the recent development of the hydrogen bomb by the Soviet Union had "outstripped the progress made in our civil defense strides to defend against it."

~The History Channel [12]

By the 1970s, the professional management of community risk shifted slightly from the civil defense model to include a so-called emergency preparedness phase of readiness. This included readiness and mitigation efforts typically focused on the "big" hazard in a given community. This hazard might be an earthquake or a volcano in the geographic "Ring of Fire" or a tornado in so-called "Tornado Alley." The nickname or cultural description often indicated these high-attention items. Unfortunately, identifying these hazards was only a minor improvement from the sole focus on the threat of nuclear fallout. Each considered community still had numerous other hazards that created risk (of various probabilities and consequences), but were often ignored.

By 1979, community leaders at all levels of government were growing concerned with the ineffectiveness of professional responses to risk.

Ironically, during the final stages of review related to the establishment of a central government organization to address these issues (which later became the Federal Emergency Management Agency [FEMA]), the United States was impacted by a major civilian nuclear accident on Three Mile Island near Harrisburg, Pennsylvania (see Figure 8.6) [13]. Even though the meltdown at Three Mile Island was not the nuclear incident long feared, it was the final nail in the proverbial coffin of both the traditional civil defense and targeted emergency preparedness approaches. Between the formal establishment of FEMA and altered political pressures, a new phase of emergency management and risk assessment was created.

Figure 8.6 The United States was heavily impacted by the response and cleanup from the Three Mile Island nuclear accident that occurred in 1979. (Source: "Report of the President's Commission on the Accident at Three Mile Island: The Need for Change.")

In Other Words...Impact of Political Decisions on the Demise of Civil Defense

In keeping civil defense funding low, Congressional leaders had little public opposition to fear. In contrast to generally widespread public participation and acceptance in the peak years of civil defense during the early stages of the Cold War, most people by this time had little faith that any government civil defense planning could lessen the impact of nuclear war. Some local communities refused outright to cooperate with Federal civil defense mandates because they did not believe [they] would be effective if a nuclear attack were to occur."

~US Department of Homeland Security [13]

This new approach focused on the development of plans and preparedness activities that address all hazards and threats faced by a given community. This process included all readiness and planning for "direction, control and warning systems which [were] common to the full range of emergencies from small isolated events to the ultimate emergency" [13]. This approach became known as the all-hazards approach, which was the first professional application that strongly attempted to both subjectively and objectively identify the probability and consequences of all threats and hazards in a given community to fully identify risk. These identifications looked for similarities and synergies between the resources needed for various risks to help create efficiency and effectiveness in community planning. This all-hazard approach coincided with a national push to formalize the professional response to risk into what became the field of emergency management [14].

This all-hazards approach remained the best practice for a long period of time, until the terrorist attacks of September 11, 2001, altered this focus to a homeland security and prevention position. This shift in risk perception and focus was shaped by the US Congress, which pushed for a large and ambitious new federal agency called the Department of Homeland Security (DHS) that would reorganize the various preparedness, public safety, and intelligence components of the federal government under a unified direction (much like FEMA's creation did in 1979) [13] (see Figure 8.7). This new agency collected all or part of 22 different federal agencies, including FEMA, Customs and Border Protection, Transportation Security Administration (TSA), US Coast Guard, and US Secret Service [15].

Figure 8.7 The US Department of Homeland Security (DHS) was established after the September 11 terrorist attacks to consolidate various governmental agencies responsible for prevention, preparedness, mitigation, response, and recovery. (Source: White House.)

As such, the focus of risk throughout the United States (and much of the developed world) strongly shifted toward prevention of and protection from terrorism. The emergency management phases previously presented as preparedness, response, recovery, and mitigation were specifically adjusted to include prevention, which was solely added due to the new threat of terrorism. Moreover, the federal government committed hundreds of millions of dollars in grants to state and local preparedness entities, leading to a significant, yet terrorism-focused surge in the resources, capabilities, equipment, and personnel available to respond in many communities.

However, by 2010, politicians and emergency managers yet again began to shift away from the heavy focus on terrorism. For practical reasons, like a depressed economy, much of the funding expired or was not renewed. Additionally, major nonterrorism or natural hazards had significant and often historic impacts on local communities. Specifically, during the first decade of activity under the guidance of DHS, Hurricane Katrina ravaged much of the Gulf Coast region of the United States, major tornadoes wiped away communities (e.g., Greensburg, Kansas), flooding continued to impact coastal regions, and earthquakes happened in regions unprepared for their effects (e.g., central Virginia). Much like the Cold War singular focus on nuclear threats, the myopic focus on terrorism during that period may (or may not) have had a significant impact on readiness for other natural hazards. It is, however, worth considering the possibility that history has strongly indicated that risk perception has a tendency to drive the general public, emergency managers, and politicians toward the perceived biggest threat, which has often led to unintended consequences from the remaining hazards.

While it is too early to definitively understand where risk evaluation and identifications are going within emergency management and communities abroad, there are some early indications that the approach will swing back toward a broader, more all-encompassing approach. Professional organizations and governmental agencies are beginning to focus on complex and comprehensive risk assessment strategies such as the threat and hazard identification and risk assessment (THIRA). This type of approach to risk assessment adds two additional layers of consideration by including context to the risk and identifying the impact to current planning standards or capabilities. For example, a low-magnitude earthquake is insignificant to most communities, while a high-magnitude earthquake is almost always catastrophic. This context could also apply to the proximity of the actual hazard or threat (i.e., the closer you are to the earthquake, the worse the effects are). Without context, these types of hazards may ultimately be overestimated or underestimated. Likewise, it is important to place contextualized risk within a broader planning framework to ensure resources and capabilities are disseminated as efficiently and effectively as possible.

DISTORTED PERCEPTIONS

A variety of research has consistently shown over the last 30 years that there are real and somewhat predictable distorted perceptions related to hazards, threats, and ultimately risk in communities of all sizes. These distortions impact local communities, regional areas, states, and countries with various applications. One of these perceptions is that low-probability events are systematically misjudged by the community and their leaders. For example, the risk from flood is often perceived as a statistically predictable event, particularly in light of government ratings such as "100-year floods" or "500-year floods" [16]. However, in reality these are statistical evaluations that are related to probability and not actuality. In the case of flooding, these terms are more appropriately understood as events that are likely to occur once every 100 or 500 years, respectively. Given this misunderstanding, community members may misunderstand that a so-called "100-year" flood can still actually occur in successive or near successive years.

Likewise, people often maintain misperceptions related to newly constructed mitigation measures such as dams and levees. Specifically, people assume that disaster situations cannot happen if these new or updated systems have been implemented [16]. In reality, newly constructed mitigation or land management systems can occasionally lead to unforeseen additional risks due to shifting pressure points within a much broader physical or ecological system. For example, the American Society of Civil Engineers (ASCE) released a report card of American infrastructure and assessed the overall US levee system a grade of "D–" because more than 85% of the nation's levees were privately built more than 50 years ago, are poorly maintained, and are not designed to adequately mitigate against current levels of community development [17]. Consequently, even new or altered levees can cause failures in nearby systems or at other levees included within the broader network of flood mitigation components.

People also alter their own view of risk based on frequency and costs related to given hazards or threats. For example, infrequent events often lead to reduced personal and community preparedness. This is particularly common in areas with low-level seismic activity. People either forget about the possibility of earthquakes or assume they will simply be minor "rumbles" that often go unfelt in a given community. For example, the 5.9 M_W earthquake in 2011 that originated in central Virginia and impacted much of the mid-Atlantic United States was surprising to many local community members and their leaders as well as national government leaders in Washington, DC, even though the area had been well documented as an active seismic area.

On the other hand, when a multiyear planning and preparedness campaign was initiated by FEMA and related partners to bring awareness to the threat of earthquakes in the midwestern United States, there was very little change in the perception of risk related to the given hazard. Even though FEMA and partners stressed the potential catastrophic risk from a major earthquake in the region, it is unlikely local residents significantly changed their risk perception and related preparedness activities since it had been more than 200 years since a major earthquake occurred. Specifically, FEMA routinely touted a major earthquake that struck the area between St. Louis and Memphis in 1811 that, if repeated, would be one of the worst disasters in recorded history. The challenge for local and regional residents is the question of proximity and timing. More recent and frequent events, including tornadoes, flooding, terrorism, and the like, have impacted their lives while an earthquake has not. This predictable pattern makes risk acceptance and action very difficult in communities even if risks are probable and would have significance.

The last major predictable influencer of risk acceptance is that of the traditional media and, more recently, social media. Traditional media sources in print, radio, and television often frame disaster events around a narrative that is compelling and interesting to those not impacted. Risk communication approaches have always known that clear and consistent messaging was critical for public acceptance and correspondingly appropriate protective actions (e.g., evacuation). However, with ever-increasing numbers of information sources available to the general public before, during, and after a disruptive event, this approach is difficult to say the least. For example, by 2015, research indicates that Americans will consume media (both traditional and social) approximately 15.5 hours per person per day. Moreover, Americans will watch approximately 11 hours of online videos per month [18]. This type of consumption leads to the quick and exponential distribution of information in a so-called "viral" process. The challenge is that the sources and speed may not always correlate, which can lead to confusing details or misinformation and ultimately a lack of awareness or acceptance of the actual threat or hazard being addressed.

In Other Words...Oprah Sways Public over Food Emergency

In 1996, at the height of the scandal about mad cow disease in the United Kingdom, a guest on Oprah Winfrey's talk show claimed that meat produced in the USA could cause bovine spongiform encephalopathy (BSE). "That just stopped me cold from eating another burger," Winfrey responded. Later, beef farmers from Texas sued Winfrey's show claiming that it was partly responsible for the steep decline in beef prices in the USA during the following months, even though the country did not have a single case of BSE. This episode demonstrates not only the power of the media and its influence on the public, but also how easily the public is swayed, particularly by fear, even in the absence of information.

~Jill McCluskey and Johan Swinnen [19]

It is important to note that the extraordinary access to information available through these various media outlets can lead to additional knowledge about hazards and risk, but not necessarily changes in perception. For example, researchers have shown that strongly supported hazard and risk reduction education sites maintained by organizations like FEMA, National Weather Service, or the US Geological Survey (USGS) can have a very small impact on risk perception [20]. Unfortunately, the correlation is so small that it is clear that other sociological or cultural factors vastly impact the much larger sways in risk perception documented after many small- and large-scale events. Moreover, this small positive correlation is often overwhelmed by an erosion of trust in risk experts due to the fact that experts do not always agree, knowledge of risk constantly changes, and distrust of governmental and bureaucratic sources is increasing [20]. For example, the USGS updated its national seismic hazards map in July 2014 after taking into account research from the 2011 earthquake off the coast of Japan and the 2011 central Virginia earthquake. This change identified parts of 16 states as having the highest risk for earthquakes [21]. The challenge is that those same earthquake risk zones were re-evaluated in 2008 to incorporate new research findings. Using the most currently available information to accurately assess risk is important, but these updates also impact risk perception significantly. The general public may receive this information as an overstatement of risk and take no action or be overly fearful and take concerted protective actions against earthquakes, but not other (more consequential) risks in their community.

BLACK SWAN EVENTS

This latter approach is of particular concern as emergency managers prepare for the future. Since the September 11 terrorist attacks in New York and Washington, DC, the world has been impacted by numerous large-scale disasters that were otherwise unexpected or unimagined prior to their occurrence. Examples of these events include Hurricane Katrina (2005), the Virginia Tech shooting (2007), the Deepwater Horizon oil spill (2010), the Fukushima meltdown (2011), Superstorm Sandy (2012), and many more (see Figure 8.8). Because these events have often resulted in a high number of casualties and significant impact to local, regional, or national economies, there is a significant amount of fear and concern by the general public that such an event may occur in their areas. Even professional financial investors have begun to create

Figure 8.8 Examples of so-called Black Swan events include Superstorm Sandy, which impacted much of the northeast United States in 2012. (Source: NASA.)

so-called "tail risk" funds to create financial hedges against losses from these types of events [22].

In Other Words...The Increased Risk of Black Swan Events

I...summarize the [Black Swan] triplet: rarity, extreme impact, and retrospective (though not prospective) predictability. A small number of Black Swans explain almost everything in our world from...the dynamics of historical events to elements of our own personal lives. Ever since we left the Pleistocene, some ten millennia ago, the effect of these Black Swans has been increasing. It started accelerating during the industrial revolution, as the world started getting more complicated, while ordinary events, the ones we study and discuss and try to predict from reading the newspapers have become increasingly inconsequential.

~Nassim Nicholas Taleb [23]

These highly unpredictable events have begun to be referred to as Black Swan events. This term was developed by futurist Nassim Nicholas Taleb as a metaphor to describe disruptive events that come as a surprise (i.e., undocumented in history, science, finance, and technology), have a major effect, and are only rationalized after the occurrence with the benefit of hindsight. This rationalized hindsight is perhaps the most significant to understand. Specifically, pre-existing personal and psychological biases make people in a given community "collectively blind" to the hazards, threats, and risks in a given community [24]. Because of the size and scope of these Black Swan events and those critical biases, the general public is becoming increasingly fixated on the occurrence of these events and the cascading consequences. For example, a study after the Fukushima nuclear meltdown found that the general acceptance of nuclear power (and associated risks) decreased significantly after the event even though the likelihood of such events still remained extremely low [25].

Unfortunately, this distorted view of Black Swan events has also begun to translate into formalized prevention and preparedness activities. For example, with significant school shooting events like the 2012 shooting at Sandy Hook Elementary School in Newtown, Connecticut, there has been a significant push by government leaders and school administrators to address this perceived risk. These actions have included targeted financial expenditures on items like cameras, metal detectors, and bulletproof dry-erase boards [26]. Likewise, many schools and communities have adopted the "run, hide, fight" active-shooter response concept and corresponding video that was produced for the City of Houston for nearly $500,000 [27]. The problem is that this level of attention to the active-shooter Black Swan event means that true risk assessments to determine actual risk are often ignored or underdeveloped. As such, preparedness planning, training (e.g., CPR), and drills (e.g., tornado) are potentially not conducted to the scope or scale that they should be to adequately prepare for the real rather than perceived risk.

In Other Words...School Safety in the Age of Black Swan Events

We're very concerned that a lot of schools are less safe today than they were before [the shooting at] Sandy Hook [Elementary School]... [because of] the heavy emphasis of the active shooter scenario which ignores other threats and that some of the training is not evidence-based and not proven to work…

~Michael Dorn [26]

As real and perceived risk clash within the general public, governmental offices, and professional emergency management community, the future of risk identification and assessment is murky. As was established from start to finish in this chapter, it is common to see that the hazards and threats that are recent, highly impactful, or otherwise grab the attention of the various community stakeholders are subject to receive the most attention, resources, equipment, and priority. Unfortunately, this process routinely has limited the effectiveness of preparing for those risks that are actually more likely and more impactful in a given area. However, if recent history is any indication, there is no reason to believe that the future will not continue to cycle through the "hot" hazard or threat unless a concerted effort is made to stabilize this undulation of approaches.

REFERENCES

1. Weiss, Daniel J. and Weidman, Jackie. (2013). "Federal Community Resilience Investments Swamped by Disaster Damages." Center for American Progress. http://www.americanprogress.org/issues/green/report/2013/06/19/67045/pound-foolish/. Accessed July 11, 2014.
2. "Risk Assessment." (n.d.). European Agency for Safety and Health and Work. https://osha.europa.eu/en/faq/risk-assessment-1/what-is-the-difference-between-a-hazard-and-a-risk. Accessed July 12, 2014.
3. "Asset Value, Threat/Hazard, Vulnerability, and Risk." (n.d.). Federal Emergency Management Agency (FEMA). http://www.fema.gov/media-library-data/20130726-1455-20490-5292/fema426ch1.pdf. Accessed July 12, 2014.
4. Ripley, Amanda. (2009). *The Unthinkable: Who Survives When Disaster Strikes and Why*. New York: Harmony.
5. Kaplan, Stanley and Garrick, B. John. (1981). "On the Quantitative Definition of Risk." *Risk Analysis* 1(1). http://josiah.berkeley.edu/2007Fall/NE275/CourseReader/3.pdf. Accessed July 12, 2014.
6. "Risk Perception." (n.d.). Michigan Tech. http://www.geo.mtu.edu/rs4hazards/links/Social-KateG/Risk%20Perception.htm. Accessed July 12, 2014.
7. "Malcolm Gladwell: 'David and Goliath.'" (n.d.) Kojo Nmandi Show Transcript. http://thekojonnamdishow.org/shows/2013-10-16/malcolm-gladwell-david-and-goliath/transcript. Accessed July 13, 2014.

8. Brenot, Jean, Bonnefous, Sylviane, and Marris, Claire. (1998). "Testing the Cultural Theory of Risk in France." *Risk Analysis* 18(6). http://www.paul-hadrien.info/backup/LSE/IS%20490/utile/cultural%20theory%20in%20France.pdf. Accessed July 13, 2014.
9. "Cultural Theory." (n.d). Triarchy Press. http://www.triarchypress.net/cultural-theory.html. Accessed July 13, 2014.
10. Ulleburg, Pal. (2003). "Personality, Attitudes, and Risk Perception as Predictors of Risky Driving Behaviors among Young Drivers." *Safety Science*. http://www.sciencedirect.com/science/article/pii/S0925753501000777. Accessed July 13, 2014.
11. "Social Cognitive Theory." (n.d.). University of Twente. http://www.utwente.nl/cw/theorieenoverzicht/Theory%20clusters/Health%20communication/Social_Cognitive_theory. Accessed July 13, 2014.
12. "First Nationwide Civil Defense Drill Held." (2014). History Channel: This Day in History. http://www.history.com/this-day-in-history/first-nationwide-civil-defense-drill-held. Accessed July 14, 2014.
13. "Civil Defense and Homeland Security: A Short History of National Preparedness Efforts." (2006). US Department of Homeland Security. http://www.training.fema.gov/EMIWeb/edu/docs/DHS%20Civil%20Defense-HS%20-%20Short%20History.pdf. Accessed July 14, 2014.
14. Crowe, Adam. (2013). *Leadership in the Open: A New Paradigm in Emergency Management*. Boca Raton, FL: CRC Press.
15. "Who Joined DHS?" (n.d.). US Department of Homeland Security. http://www.dhs.gov/who-joined-dhs. Accessed July 15, 2014.
16. Wilson, Sean et al. (2007). "The Lack of Disaster Preparedness by the Public and Its Effect on Communities." *The Internet Journal of Rescue and Disaster Medicine* 7(2). http://ispub.com/IJRDM/7/2/11721. Accessed July 16, 2014.
17. Jefferson, Anne. (2011). "Levees and the Illusion of Flood Control." *The World of Geology and Earth Science*. http://all-geo.org/highlyallochthonous/2011/05/levees-and-the-illusion-of-flood-control. Accessed July 17, 2014.
18. "Media Consumption to Average 15.5 Hours a Day by 2015." (2013). University of Southern California Marshall School of Business. http://www.sciencedaily.com/releases/2013/10/131030111316.htm. Accessed July 20, 2014.
19. McCluskey, Jill and Swinnen, Johan. (2011). "The Media and Food-Risk Perceptions." *Science and Society*. http://www.ncbi.nlm.nih.gov/pmc/articles/PMC3128959. Accessed July 17, 2014.
20. Sjoberg, Lennart. (1999). "Risk Perception by the Public and Experts: A Dilemma in Risk Management." *Human Ecology Review*. http://ajph.humanecologyreview.org/pastissues/her62/62sjoberg.pdf. Accessed July 20, 2014.
21. Borenstein, Seth. (2014). "Risk of Earthquake Increased for One-Third of United States." *ABC News*. http://abcnews.go.com/Technology/wireStory/risk-earthquake-increased-half-us-24605902. Accessed July 20, 2014.
22. Monro, Alex. (2011). "Investors Seek Tail Risk Funds to Cover 'Black Swan' Events." Risk.net. http://www.risk.net/credit/feature/2024521/investors-seek-tail-risk-funds-cover-black-swan-events. Accessed July 20, 2014.

23. Taleb, Nassim Nicholas. (2007). "The Black Swan: The Impact of the Highly Improbable." *New York Times.* http://www.nytimes.com/2007/04/22/books/chapters/0422-1st-tale.html?pagewanted=all&_r=0. Accessed July 20, 2014.
24. Taleb, Nassim Nicholas. (2010). *The Black Swan: The Impact of the Highly Improbable,* 2nd ed. New York: Random House.
25. Huang, Lei et al. (2013). "Effect of the Fukushima Nuclear Accident on the Risk Perception of Residents near a Nuclear Power Plant in China." *Proceedings of the National Academy of Sciences.* http://www.ncbi.nlm.nih.gov/pmc/articles/PMC3856800/. Accessed July 21, 2014.
26. McKay, Jim. (2014). "Are Schools Focusing Too Much on Active Shooter Scenarios?" *Emergency Management Magazine.* http://www.emergencymgmt.com/training/Are-Schools-Focusing-Too-Much-Active-Shooter.html. Accessed July 21, 2014.
27. Pinkerton, James. (2012). "Homeland Security Films Are a Big Hit, But Costly." *Houston Chronicle.* http://www.chron.com/news/houston-texas/article/Homeland-Security-films-are-a-big-hit-but-costly-3799862.php. Accessed July 21, 2014.

9

Risk Imbalance

> Think of the terrorist attack of September 11, 2001: had the risk been reasonably conceivable on September 10, it would not have happened. If such a possibility were deemed worthy of attention, fighter planes would have circled the sky above the twin towers, airplanes would have had locked bulletproof doors and the attack would not have taken place, period.
>
> ~Nassim Nicholas Taleb [1]

GUERILLA FIGHTING, GUERILLA RISKS

History is filled with examples of highly impactful and disruptive events. Revolutionary wars, civil wars, and world wars have impacted the world for millennia. For the vast majority of history, military conflict has been fought between two large government states in a formalized approach where each side amasses the most resources possible in an attempt to overwhelm the other side. In each case, war has leveraged an astronomical amount of resources and typically has represented a significant percentage of the capabilities of the nation-state. For example, the US government has traditionally spent 1%–4% of the gross domestic product (GDP) on military defense and preparedness during peacetime, but this figure jumped to 41% during World War II [2]. In all cases, the results of war were large scale and tragic to areas beyond the location of simply fighting or military skirmishes. These high-impact events were always the result of resource-intensive or high-input events. In many ways this balance of input and output was predictable and understood as a natural balance.

The impact of an event could not exceed the input of resources—or so the perception went.

However, this process changed significantly after World War II as traditional war was no longer fought. With the rise of guerrilla fighters during the Vietnam War, this balance shifted significantly. No longer was war defined as the high-input, high-impact event (see Figure 9.1). Under these circumstances, smaller, decentralized forces maintained equal impacts, but with significantly fewer resources required to be submitted. The application of guerrilla fighting has become standard since that time, particularly in developing countries. Moreover, it was the dominant approach that Islamic terrorist groups used in Afghanistan and Iraq for much of the first decade of the twenty-first century. These low-input, high-impact approaches have traditionally been very hard for traditional, resource-intensive organizations (e.g., armies of developing countries) to manage.

This same assumption has often been applied to emergency management and risk assessments. Natural hazards are easy to understand as the related physical size is often as impressive as the impacts they create.

Figure 9.1 During the Vietnam War, smaller and decentralized forces maintained equal impacts with significantly fewer resources needed as compared to previous global conflicts. (Source: US Marine Corps.)

For example, when Hurricane Sandy approached the northeastern coast of the United States, the size of the storm was more than 500 miles across, which is roughly twice the size of Texas [3]. Given the size and scope of the storm, it was easy to understand how it would correlate to the scale of the impact seen in New York City and beyond. Likewise, wildfires that burn millions of acres and generate fire tornadoes are equally grand in scope and understandably generate significant destruction and fatalities [4]. International reinsurance provider Swiss Re even released a report projecting $20 billion in property damage and thousands of fatalities from an EF-5 tornado (similar to that which impacted Joplin, Missouri, in 2011) impacting a large urban area such as Chicago or Atlanta [5]. Much like the discussed war, these types of natural disasters are all clearly events that have significant impact only when an equally significant amount of energy and size is input.

TERRORISM AND ACTIVE-SHOOTER EVENTS

However, this process does not always correlate to human-caused disasters like terrorism. The modern concept of terrorism was established in 1793 by Maximilien Robespierre during the so-called "Reign of Terror" following the French Revolution (see Figure 9.2). Specifically, Robespierre justified his means of overthrowing the existing form of government when he said, "Subdue by terror the enemies of liberty, and you will be right, as founders of the Republic" [6]. This sentiment laid the foundation for state terrorism intended to overthrow government forms throughout the world. These types of terrorist groups have arisen throughout the world and include examples such as the Irish Republican Army (IRA) and Palestine Liberation Organization (PLO) [6]. This form of terrorism took an organized, relatively large group of people willing to put forward rebellion to accomplish the goals. Again, state-level terrorism is very similar to the traditional war model and the natural disasters discussed earlier that only had significant results when significant resources and people were injected into the process.

However, non-state-level terrorism has also developed over the last two decades and utilizes a more decentralized model. For example, on September 11, 2001, it took only 19 Islamic militants associated with the extremist (yet decentralized) al-Qaeda group to hijack four airliners in US airspace and crash them into both World Trade Center towers and the Pentagon. These minimal resources ultimately collapsed both towers

Figure 9.2 The modern concept of terrorism was established in 1793 by Maximilien Robespierre during the so-called "Reign of Terror" following the French Revolution. (Source: portrait of Maximilien de Robespierre held at the Carnavalet Museum, Paris.)

and led to the deaths of almost 3,000 people, including 400 police officers and firefighters [7]. This event and the global "discovery" of decentralized terrorism (e.g., al-Qaeda) started a sequence of terrorist events and threats throughout the world. For example, by 2005, four suicide bombers had struck London, killing 52 people and injuring more than 770 [8]. Other missed opportunities included the so-called shoe bomber in 2001 and the underwear bomber in 2009, who tried to blow up airplanes full of people traveling over the United States (see Figure 9.3). In both cases, hundreds of people would have died if these single individuals had succeeded. Overall, unlike traditional war approaches and the state-level terrorism, these events mark a significant shift toward highly impactful events that did not leverage significant resources to accomplish the results.

Likewise, the Boston Marathon bombing in 2013 was extraordinarily impactful not only to the local community, but also to the American culture and the competitive running community throughout the world.

Figure 9.3 The so-called shoe bomber represents another example of a decentralized approach to terrorism that required minimal input of resources, but could have had significant impact. (Source: US federal government.)

Specifically, on April 15 the Tsarnaev brothers detonated two bombs near the finish line of the marathon which ultimately killed three spectators and wounded more than 260 others. Over the course of 4 days, local, state, and federal police agencies conducted a large-scale investigation including nearly 24 hours of an intense manhunt which included the shooting death of a local law enforcement officer and the death of one of the two brothers [9]. According to one risk management company, this process cost Boston approximately $333 million due to lost wages, lost retail sales, and infrastructure damage [10]. Considering the miniscule costs of the bombs the Tsarnaev brothers built, this is an astronomical imbalance of resources versus results.

These unbalanced results of small-scale terrorism were also not limited to the number of fatalities or impacts to buildings. The cascading cultural and economic impacts were tremendous as well. For example, the 9/11 terrorist attacks resulted in $10 billion to $13 billion in infrastructure damage, $40 billion in government emergency funds for changes to air travel security, $17 billion in lost wages from the direct loss of 83,000 jobs, and $40 billion in insurance and restoration costs [11]. These financial costs only begin to address some of the broader cultural impacts (see Figure 9.4). For example, security for air travel has increased significantly with far more inspectors, the implementation of an air marshal program, secondary screenings, "no fly" lists, and other individually impactful processes like the removing of shoes for security checks and limits to the size and quantity of liquids that can be brought on planes. These concrete and abstract examples simply reinforce this new model of

Figure 9.4 The 2001 attacks on the World Trade Center caused astronomical impacts to infrastructure, air travel, and personal security. (Source: FEMA/Andrea Booher.)

high-impact events that ultimately require very little input to accomplish the extreme results.

In Other Words...The True Costs of 9/11

Al Qaeda spent roughly half a million dollars to destroy the World Trade Center and cripple the Pentagon. What has been the cost to the United States? In a survey of estimates by *The New York Times*, the answer is $3.3 trillion; or about $7 million for every dollar Al Qaeda spent planning and executing the attacks. While not all of the costs have been borne by the government—and some are still to come—this total equals one-fifth of the current national [American] debt.

~Shan Carter and Amanda Cox [11]

Likewise, active-shooter events are showing similar patterns of extreme impact from minimal resources. Arguably the first and certainly the most notable active-shooter event occurred in April 1999, when two students walked into Columbine High School in Littleton, Colorado, and killed 12 students and one teacher. In this example, the local SWAT team entered the school 47 minutes after the shooting began and declared the school safe after 5 hours [12]. Unfortunately, since the Columbine shooting there have been between 84 and 110 active-shooter events, depending on the source and approach, with a steady increase from year to year since that event [13]. Similar high-profile active-shooter events include the 12 fatalities and 58 injuries in just under 15 minutes during the 2012 Aurora, Colorado, movie theater shooting and the 32 fatalities and 17 additional wounded in less than 11 minutes during the 2007 Virginia Tech shooting [14].

These noteworthy events are not exceptions to the rule, but rather a normative example of the speed and impact of active-shooter events. For example, one academic study of 35 active-shooter events found that the average active-shooter event lasted 12 minutes; 37% lasted less than 5 minutes. Moreover, 66% of the shooters either committed suicide or were killed in the response [15]. To this day, all of these active-shooter events have significantly impacted not only their locally impacted community, but also broader cultural considerations related to school safety, mental health issues, and the protection of vulnerable populations. But much like the non-state-terrorism trend discussed earlier, where a very few number of perpetrators have a significant impact, active-shooter events have similar imbalances. Typically, one shooter with a handful of weapons can kill or injure dozens of individuals (or more) in a manner of minutes and shake the foundations of the community for many years.

In Other Words...Tyranny of the Singular

We are...subject to the tyranny of the singular, the accidental, the unseen and the unpredicted.... Lone individuals can alter the course of history. People kill each other every day without much physical exertion...and at the same time, we have become ever more interdependent.

~Nassim Nicholas Taleb [1]

BLACK SWAN EVENTS

In addition to the small input but high consequence of terrorism and active-shooter events, there is another type of risk imbalance that is of growing importance in recent history and certainly as emergency managers prepare for the future. Specifically, as briefly discussed in Chapter 8, the rise (or identification) of so-called Black Swan events has significantly changed how the public perceives certain risks and how emergency managers must begin to prepare for them. As described in the previous chapter, a noted scholar named Nassim Nicholas Taleb first used the term "Black Swan" to identify those incidents that are extremely rare, highly impactful, and only retrospectively predictable [1]. Examples of Black Swan events are sprinkled throughout history, but increased in frequency after the Industrial Revolution and include the 9/11 terrorist attacks, Hurricane Katrina, Superstorm Sandy, the *Challenger* explosion, the Fukushima meltdown, and many more [16].

Because of this unique set of circumstances, standardized statistical analysis and predictability processes are skewed. This is emphasized by the fact that Black Swan events are inherently outlier events that neither add nor take away from statistical evaluations. Taleb succinctly addresses the outlier nature by stating that "nothing in the past can convincingly point to its possibility" [16]. However, given the extreme impact of Black Swan events and regardless of the outlier statistical status, human beings always attempt to find retrospective explanations for these events. The personal and professional explanations of Black Swan events are often varied and contradictory, which ultimately leads to further confusion about the event and ultimately a misappropriation of resources to address an event that may never happen again. Unfortunately, these events are fundamentally not predictable and should be treated as such.

Thus far, the discussion of Black Swan events has solely focused on the negative impact of emergencies and disasters. It is important to acknowledge that Black Swan events by definition can be equally positive. For example, the rise of the Internet would meet the definition of a Black Swan event. Specifically, it was created as an outlier form of government communication redundancy, has been revolutionary in its global impact, and ultimately can only be partially explained by looking retrospectively. Taleb describes these positive Black Swans as "unlikely bouts of extreme good fortune" [17]. The only significant difference between the two types of Black Swan events is that positive events tend to be much slower and not realized (or appreciated) until they are fully formed, while negative Black Swan events strike hard and fast [1].

FINDING THE BLACK SWANS

Somewhat ironically, the traditional media, politicians, citizens, and even emergency managers are seeking out ways to identify and ultimately prepare for the next Black Swan event. For example, military futurist Dr. Andrew F. Krepinevich, Jr., wrote in his book, *7 Deadly Scenarios: A Military Futurist Explores War in the 21st Century*, that there were seven Black Swan events that would impact the world before 2030. Specifically, Krepinevich identified a severe pandemic, rapidly increasing global warming, European economic collapse, a democratic Chinese governance, a reformed Iran, a global nuclear attack, and a US global disengagement [18] (see Figure 9.5). While these particular Black Swan events are focused on global political and military risk, emergency managers have similar, yet unofficial lists that include hazards and threats such as catastrophic earthquakes, near-Earth objects (NEOs), nuclear meltdowns, climate change, and many more.

The first of these potential Black Swan events is a massive and catastrophic earthquake. For example, in 2006, the US Federal Emergency

Figure 9.5 Black Swan events like the 1918 pandemic flu could wreak havoc on global security and safety. (Source: *St. Louis Post Dispatch*.)

Management Agency (FEMA) provided funding to the Central United States Earthquake Consortium (CUSEC) to create the New Madrid Seismic Zone (NMSZ) Planning Project. This NMSZ Planning Project was a very large multistate response planning effort as it drew upon lessons learned from other seismic events and current scientific thinking by utilizing the planning resources of the states of Alabama, Arkansas, Illinois, Indiana, Kentucky, Mississippi, Missouri, and Tennessee (which were most at risk from this particular fault line). The justification for this comprehensive action was a series of earthquakes in 1811 and 1812 that were estimated to be between 7.0 and 8.0 M_W [19]. Likewise, the US Geological Survey (USGS) reports that the New Madrid fault line generates nearly 200 low levels of magnitude earthquakes annually, which justify its classification of the area as highly seismic [20].

Geologists and seismologists continue to maintain that the NMSZ located between St. Louis, Missouri, and Memphis, Tennessee, is highly conducive to movement and the potential damage area is more widespread than other earthquake-prone areas. They project that this area is impacted by a 400- to 500-year cycle of earthquakes (although this is highly debatable) and would cause as much as $70 billion in damages with compounded infrastructure issues in the area [21]. The real challenge is that this area has not had a major earthquake in nearly 200 years. Moreover, even if the scientific projections are correct about the seismic cycle in this area, there are another 200 years until the risk of a catastrophic event would occur.

Meanwhile, cities in the New Madrid fault area have routinely experienced other hazards like flooding and tornadoes. For example, St. Louis, Missouri, has experienced 39 tornadoes since 1950, including EF4 tornadoes in 1959, 1967, and 2011, which resulted in 13 deaths and nearly 400 injuries [22]. Similarly, more than 40 major floods have occurred along the Mississippi River basin (which overlaps with much of the NMSZ) since 1811 [23]. While there is legitimate seismic risk from earthquakes to those areas included in the NMSZ, it must be more appropriately considered when assessing risk by comparing it to other natural and human-caused hazards that might impact the area. Unfortunately, the Black Swan status of a major earthquake in the NMSZ gets an overwhelming amount of attention and resources which could be used as part of an all-hazard approach to prioritize resources, but more than likely creates an imbalance of risk for citizens and responders alike.

Like the possibility of a New Madrid earthquake, another Black Swan often hyped by traditional media outlets and some community groups is the so-called "supervolcano." According to the USGS, a supervolcano implies a volcanic center that registers no less than an eight on the Volcano Explosivity Index (VEI) with measured deposits of greater than 1,000 cubic kilometers (240 cubic miles) from an eruption. However, the last time VEI 8 eruptions occurred was tens of thousands to millions of years ago. For example, supervolcanoes last occurred approximately 2 million years ago near Yellowstone National Park and Toba (Indonesia). The most recent supervolcanic eruption occurred 27,000 years ago near Taupo in New Zealand [24]. However, much like the catastrophic earthquakes mentioned earlier, there are some real (and perhaps some fictitious) risks from supervolcanoes. While estimates vary, one media outlet reported that a supervolcano in the Yellowstone basin would immediately kill 87,000 people, spread ash across a majority of the country, and cause as much as $3 trillion in damage [25]. While even more unlikely than the catastrophic earthquake, this Black Swan event still receives some attention from media and the general public and in turn may cause an additional burden on emergency managers as they assess what risks to address and how to explain them to the general public.

In Other Words...Will Another Supervolcano Occur?

Although it is possible, scientists are not convinced that there will ever be another catastrophic eruption at Yellowstone. Given Yellowstone's past history, the yearly probability of another...[massive] eruption could be calculated as 1 in 730,000 or 0.00014%. However, this number is based simply on averaging the two intervals between the three major past eruptions at Yellowstone—this is hardly enough to make a critical judgment. This probability is roughly similar to that of a large asteroid hitting the Earth. Moreover, catastrophic geologic events are neither regular nor predictable.

~USGS [24]

Although not technically a Black Swan event (as it can be forecasted), another risk imbalance that has made periodic splashes in the media is so-called near-Earth objects. According to NASA, NEOs are comets and asteroids that have been pushed by the gravitational attraction of planets into orbits that allow them to pass near Earth [26] (see Figure 9.6).

Figure 9.6 Near-Earth Objects (NEOs) are comets and asteroids that have been pushed by the gravitational attraction of planets into orbits that allow them to pass near Earth. (Source: NASA.)

While NASA studies NEOs to better understand the physical and chemical makeup of space material, some scientists have projected that these NEOs pose a significant threat to Earth, with some attributing the extinction of the dinosaurs to the impact of a large meteor on Earth [27]. Commercial movies like *Armageddon* and *Deep Impact* play to this fear and theoretical risk to Earth and all her inhabitants. However, of the nearly 10,000 NEOs that NASA has identified and assessed risk for, the most likely NEO to impact Earth is only given a 0.00001% probability of happening [28].

However, even with that astronomically small probability of a NEO impact to Earth, the US federal government provided $20.4 million in 2010 to expand the Near-Earth Orbit Observation Program to "improve and increase its efforts to detect Earth approaching asteroids and comets that may...become potential impact hazards to Earth...and provide information for study of potential hazard mitigation techniques" [29]. Moreover, Congress mandated that NASA discover at least 90% of

1-kilometer-sized NEOs. This budget was later increased by another $20 million with an NEO disaster response plan and exercise scenario developed and conducted by FEMA [30]. Likewise, the United Nation's Scientific and Technical Subcommittee of the Committee on the Peaceful Uses of Space helped establish an International Asteroid Warning Network to collect and share information about potentially hazardous NEOs and facilitate a "spacecraft mission intended to deflect [an NEO] from its collision course with Earth" [31]. This is an incredibly complex planning response for something with so little potential impact and only a proposed mitigation strategy. Much like the previous Black Swan events, an NEO impact to Earth is possible, but truly an event of distorted risk.

POLITICAL RESPONSE TO RISK IMBALANCE

As discussed in Chapter 7, disasters can be highly politicized at all phases of disaster management, including preparedness, response, recovery, and mitigation. These are complex enough for legitimate risks that impact a given community, but become astronomically more delicate for catastrophic events that happen or may happen in a given community. This is particularly true at the federal level among the presidents, prime ministers, Congress, and Parliament. For example, every American president since Lyndon B. Johnson has had to take quick and decisive political actions immediately following catastrophic or highly publicized events such as the September 11 terrorist attacks [32,33].

In Other Words...Tough Political Decisions Related to Black Swan Events

For government officials, few problems are tougher than deciding how best to head off rare, but potentially devastating, event risks. Do too much and you impose unreasonable costs and hurt the economy in response to a problem that might not happen for centuries. Do too little and you add to a list of unheeded disaster warnings that include the risk of storm surges in New Orleans and tsunamis in the Indian Ocean.

~Evan Halper [34]

Political sensitivity has continued to be impactful to current world leaders. For example, US President Barack Obama received a 10% positive increase in his public approval ratings after the death of Osama Bin Laden and the 2011 outbreak of tornadoes in Missouri and Alabama, but received a significant drop after the extended response to the BP oil spill in the summer of 2010 [35,36] (see Figure 9.7). Likewise, South Korean President Park Geun-hye maintained significant political pressure and 4 weeks of declining approval ratings after the national response to the sinking of a ferry that killed nearly 300 people in April 2014 [37]. While some observers would merely designate these political undulations as part of the life of an elected official, the ebb and flow must be more carefully considered as political actions almost always immediately follow major catastrophic events—whether predictable or not.

One of the most significant examples of this impulsive political behavior after disasters is the creation of the US Department of Homeland Security (DHS) following the terrorist attacks of September 11, 2001. Eleven days after the terrorist attacks, Pennsylvania Governor Tom Ridge was appointed by US President George W. Bush to serve as the first director of the office of homeland security in the White House and to oversee the "comprehensive national strategy to safeguard the country against terrorism and response to any future attacks" [38]. By the next

Figure 9.7 US President Barack Obama received increases and decreases in public perception ratings after his leadership during responses to various disasters and Black Swan events. (Source: FEMA/Aaron Skolnik.)

November, the US Congress passed the Homeland Security Act, which established the Department of Homeland Security (DHS) as a stand-alone, cabinet-level department which ultimately opened its doors March 2003 as a collection of 22 different pre-9/11 units, agencies, or departments (e.g., the Transportation Security Administration and FEMA) [38]. This process included an initial $19.5 billion budget in 2002 that was nearly doubled to $37.7 billion in 2003. This budget has continued to climb since then with a total budgetary authority of nearly $61 billion in 2015 [39,40]. All said, this is an astronomical response to something that previously was almost unheard of in the United States. That is not to identify this is an inappropriate response, but rather to simply consider the imbalance of response to risk.

In addition to the massive funding required to establish the DHS, the Provide Appropriate Tools Required to Intercept and Obstruct Terrorism (PATRIOT) Act was signed into law by President George W. Bush on October 26, 2011, a mere 6 weeks after the initial act of terrorism. The PATRIOT Act "broadened the government's authority to collect domestic records and surveil its citizens…without evidence of wrongdoing" [41]. This significant shift in authority and change in oversight was thought to be a small sacrifice of personal freedoms for the protection of the greater good. However, it quickly became vitriolic among citizens and politicians alike with groups like the American Civil Liberties Union (ACLU) bringing legal suits against the government. Interestingly, the PATRIOT Act was reauthorized by President George W. Bush in 2006 and again by President Barack Obama in 2011 [41].

While the changes following the 9/11 terrorist attacks are significant, it is not the only political response worth discussing. For example, while there have been nearly 100 active-shooter events (see earlier) in the last decade, there is no consensus among politicians, emergency managers, or the general public as to the root cause of these events. Some individuals have argued that active-shooter events are only possible because of the ease of access to guns of various caliber, size, and capacity. Simultaneously, others have argued that active-shooter events are a direct result of a broken national mental health care system that cannot care for those identified with conditions that might ultimately lead to harming self or community. Lastly, others have argued that these events are unavoidable and school systems should be structurally and nonstructurally modified to prepare to respond to these issues.

For example, after the 2012 school shooting at Sandy Hook Elementary school in Newtown, Connecticut, there was a surge in public discourse regarding the need for gun control to prevent similar events from occurring in the future. Specifically, President Barack Obama stated at a vigil for the event, "Are we prepared to say that such violence visited on our children year after year is somehow the price of our freedom?" [42]. Advocates—political and public—jumped on these and similar comments identifying the United States as having a higher rate of gun deaths than many other developing countries. In contrast, advocates for the US Second Amendment staunchly pointed out their constitutional (if not fundamental) right to "keep and bear arms," with advocacy built around the concept that guns are not inherently lethal, but only become so when the intention of humans leverages them as so [42]. The problem with leveraging a disaster or disruptive event (as in this case the Newtown shooting) is that this issue is broader than the disaster itself. The issues of gun control are very contentious and ultimately may or may not decrease risk from active-shooter events, but may be opportunistically leveraged by politicians, which may undermine an emergency manager's attempt to efficiently and effectively manage risks in a community now and into the future.

Likewise, other risk managers, media members, and citizens have expressed that failures in the mental health system are the primary cause of active-shooter events. Typically, proponents of this root cause identify lack of psychological training, available time, or availability of functional support programs often allowing those with mental illness and a desire to cause harm to fall through the proverbial cracks of government and health care protection systems. In support of this concept, the US Department of Defense (DoD) modified mental health care support services in the aftermath of the active-shooting/terrorism event perpetrated by Major Nidal Hasan at Ft. Hood, Texas, in 2009 (see Figure 9.8). Specifically, the DoD created a Ft. Hood Behavioral Health Plan that incorporated a whole-community approach for long-term behavioral health care and to reduce risk from future similar events [43]. Conversely, both the US Federal Bureau of Investigation (FBI) and the Police Executive Research Forum (PERF) have stated that while active shooters may have mental health issues prior to the event, their psychological condition is an "indeterminate" causal factor [44,45]. However, much like the previously discussed impact of gun control, mental health is a political "hot potato." Even after the Newtown shooting, when President Obama committed $100 million to significant public approval, there was still significant political pushback calling the contribution a "drop in a bucket" and pointing out that individual states had contributed more than twice that [46]. The issue of improving

Figure 9.8 The US Department of Defense (DoD) modified mental health care support services in the aftermath of the active shooting/terrorism event perpetrated by Major Nidal Hasan at Ft. Hood, Texas, in 2009. (Source: US Department of Defense.)

support systems for mental health are opportunistically addressed after disasters like active-shooter events, but ultimately may again hinder how risk is actually handled in communities due to distortion and distraction.

In Other Words...The Role of Mental Health in Active-Shooter Events

Though mental illness is a common factor among many active shooters, its functional role in causing the massacre is indeterminate according to FBI analysis. Very few of the shooters in cases analyzed by the FBI had previous arrests for violent crimes, though many had encountered a significant emotional hardship prior to the attack such as "loss of significant relationships, changes in financial status, loss of job...and/or feelings of humiliation or rejection on the part of the shooter."

~Public Intelligence Report [47]

Lastly, other people put the blame for active-shooter events—particularly those in schools—on the facility preparedness. To address this concern, many school districts have implemented a wide spectrum of response and/or mitigation strategies such as metal detectors and bulletproof instruction boards. Other places have crowdsourced information from student, faculty, and parents about current conditions of the school grounds to identify things like broken locks and windows. Unfortunately, the increased focus on active-shooter threats in schools has also made them like a "fortress" that neglects other hazards that may impact the institution [48]. This is a critical distinction as FBI statistics have shown that only 17% of the approximately 100 active-shooter events in the last 12 years occurred in schools [15]. Consequently, over that period of time, active-shooter events only occurred in 0.05% of the nearly 40,000 public and private secondary schools in the United States [49]. Without the distorted perception of risk related to active shooters, the commitment and dedication might not otherwise occur. While some community leaders and emergency managers might argue that the resources dedicated to active shooters have secondary or parallel uses for other risk, the focus, attention, planning, and preparedness may be eroded by this continued approach to what ultimately is a Black Swan type of event at a local school.

REFERENCES

1. Taleb, Nassim Nicholas. (2010). *The Black Swan: The Impact of the Highly Improbable*, 2nd ed. New York: Random House.
2. "A Century of Defense Spending." (n.d.). US Government Spending. http://www.usgovernmentspending.com/defense_spending. Accessed July 22, 2014.
3. Duke, Alan. (2012). "Superstorm Sandy Breaks Records." CNN. http://www.cnn.com/2012/10/30/us/sandy-records. Accessed July 22, 2014.
4. Burt, Christopher C. (2011). "The Worst Wildfires in World History." *Weather Underground*. http://www.wunderground.com/blog/weatherhistorian/the-worst-wild-fires-in-world-history. Accessed July 22, 2014.
5. Rice, Doyle. (2014). "What If a Tornado Hit a Major U.S. City?" *USA Today*. http://www.usatoday.com/story/weather/2014/07/22/tornado-cities-swiss-re-chicago/13000993. Accessed July 23, 2014.
6. Zalman, Amy. (n.d.). "1st Century BCE–13th Century: Terrorism in the Pre-Modern World." About.com. http://terrorism.about.com/od/whatisterroris1/p/Terrorism.htm. Accessed July 23, 2014.
7. "9/11 Attacks." (2014). History Channel. http://www.history.com/topics/9-11-attacks. Accessed July 23, 2014.

8. "7 July Bombings." (2005). *BBC News*. http://news.bbc.co.uk/2/shared/spl/hi/uk/05/london_blasts/what_happened/html/. Accessed July 23, 2014.
9. "Boston Marathon Bombing." (n.d.). History Channel. http://www.history.com/topics/boston-marathon-bombings. Accessed July 23, 2014.
10. Rinaldi, Jessica. (2013). "Adding Up the Financial Costs of the Boston Bombing." *NBC News*. http://usnews.nbcnews.com/_news/2013/04/29/17975443-adding-up-the-financial-costs-of-the-boston-bombings. Accessed July 23, 2014.
11. Carter, Shan and Knox, Amanda. (n.d). "One 9/11 Tally: $3.3 Trillion." *The New York Times*. http://www.nytimes.com/interactive/2011/09/08/us/sept-11-reckoning/cost-graphic.html?_r=0. Accessed July 23, 2014.
12. "Columbine High School Shooting Fast Facts." (2014). CNN. http://www.cnn.com/2013/09/18/us/columbine-high-school-shootings-fast-facts/. Accessed July 24, 2014.
13. Blair, J. Pete, Martindale, M. Hunter, and Nichols, Terry. (2014). "Active Shooter Events 2000–2012." *FBI Law Enforcement Bulletin*. http://leb.fbi.gov/2014/january/active-shooter-events-from-2000-to-2012. Accessed July 24, 2014.
14. "Special Analysis: The Active Shooter Threat." (2012). *MSA Worldview*. http://www.msasecurity.net/Portals/91068/docs/MSA%20Special%20Analysis%20-%20Active%20Shooter%20Threat%208.20.12.pdf. Accessed July 24, 2014.
15. "Active Shooter Statistics." (n.d.). FBI Critical Incident Response Group. http://www.fbi.gov/about-us/cirg/active-shooter-and-mass-casualty-incidents/active-shooter-statistics. Accessed July 24, 2014.
16. Taleb, Nassim Nicholas. (2007). "The Black Swan: The Impact of the Highly Improbable." *The New York Times*. http://www.nytimes.com/2007/04/22/books/chapters/0422-1st-tale.html?pagewanted=all&_r=1&. Accessed July 25, 2014.
17. Moyer, J. D. (2010). "Exposing Yourself to Black Swan Events." J.D. Moyer Blog. http://jdmoyer.com/2010/05/20/exposing-yourself-to-positive-black-swans/. Accessed July 25, 2014.
18. Galdorisi, George. (2013). "Global Trends 2013: Black Swans." Defense Media Network. http://www.defensemedianetwork.com/stories/global-trends-2030-black-swans/. Accessed July 25, 2014.
19. "CUSEC New Madrid Seismic Zone Catastrophic Planning Project." (n.d.). Central United States Earthquake Consortium. http://www.cusec.org/plans-a-programs/multi-state-planning/156-cusec-new-madrid-seismic-zone-catastrophic-planning-project.html. Accessed July 25, 2014.
20. Oskin, Becky. (2014). "Thanks to Active New Madrid Fault, Midwest Earthquake Risk Still Looms." The Weather Channel. http://www.weather.com/news/science/thanks-active-new-madrid-fault-zone-midwest-earthquake-risk-still-looms-20140124. Accessed July 25, 2014.
21. "New Madrid Seismic Zone Catastrophic Planning." (2008). Federal Emergency Management Agency (FEMA). http://www.cusec.org/documents/cusec/nmszplanning.pdf. Accessed July 25, 2014.

22. "Tornadoes in St. Louis, Mo.—1950–2001." The Weather Channel. http://www.weather.com/outlook/weather-news/severe-weather/articles/tornadoes-by-month-saint-louis_2010-03-25. Accessed July 26, 2014.
23. Trotter, Paul S. et al. (1998). "Floods in the Lower Mississippi: A Historical Economic Overview." National Weather Service. http://www.srh.noaa.gov/topics/attach/html/ssd98-9.htm. Accessed July 26, 2014.
24. "Questions about Supervolcanos." (2014). USGS. http://volcanoes.usgs.gov/volcanoes/yellowstone/yellowstone_sub_page_49.html#supervolcano. Accessed July 26, 2014.
25. Lemas, Matt. (2014). "If This Supervolcano Erupts, Two-Thirds of America Will Be Screwed." http://www.ryot.org/supervolcano-yellowstone-volcano-will-it-erupt/765673. Accessed July 26, 2014.
26. "Near Earth Object Program." (2014). National Aeronautics and Space Administration (NASA). http://neo.jpl.nasa.gov/neo/. Accessed July 27, 2014.
27. Carrill, Luis Fernandez. (2012). "The Evolution of Near Earth Objects Risk Perception." *The Space Review*. http://www.thespacereview.com/article/2080/1. Accessed July 27, 2014.
28. "Sentry Risk Table." (2014). NASA. http://neo.jpl.nasa.gov/risk/. Accessed July 27, 2014.
29. Johnson, Lindley. (2010). "Potentially Hazardous Asteroid Workshop." NASA. http:/neo.jpl.nasa.gov/neo/2011_AG5_LN_intro_wksp.pdf. Accessed July 27, 2014.
30. "NASA's Search for Asteroids to Help Protect Earth." (2014). NASA. http://www.nasa.gov/content/nasa-s-search-for-asteroids-to-help-protect-earth/#.U9Wm0PldWSo. Accessed July 27, 2014.
31. David, Leonard. (2014). "United Nations Takes Aim at Asteroid Threat to Earth." Space.com. http://www.space.com/24671-asteroid-threat-united-nations.html. Accessed July 27, 2014.
32. Zelizer, Julian E. (2011). "When Natural Disasters Become Political Disasters." CNN. http://www.cnn.com/2011/OPINION/08/27/zelizer.natural.disaster.politics/. Accessed July 28, 2014.
33. Fletcher, Michael A. and Morin, Richard. (2005). "Bush's Approval Rating Drops to New Low in Wake of Storm." *Washington Post*. http://www.washingtonpost.com/wp-dyn/content/article/2005/09/12/AR2005091200668.html. Accessed July 28, 2014.
34. Halper, Evan. (2014). "Concern Grows over Possibility of Massive Power Surge." *Los Angeles Times*. http://articles.latimes.com/2014/feb/01/nation/la-na-power-surge-20140202#axzz2sGtEhLqZ. Accessed July 27, 2014.
35. Smith, Allan. (2014). "Here Are the Triumphs and Disasters That Coincide with President Obama's Approval Ratings." *Business Insider*. http://www.businessinsider.my/triumphs-and-disasters-peaks-and-valleys-obama-2014-7/#.U9fGA_ldWSp. Accessed July 29, 2014.
36. Newport, Frank. (2010). "Obama Receives 44% Approval on Oil Spill While BP Gets 16%." *Gallup Politics*. http://www.gallup.com/poll/140957/obama-receives-approval-oil-spill-gets.aspx. Accessed July 29, 2014.

37. Sang-Hun, Choe. (2014). "South Korean Leader Dismisses More Aides after Ferry Disaster." *New York Times*. http://www.nytimes.com/2014/05/23/world/asia/south-korean-leader-dismisses-more-aides-after-ferry-disaster.html?_r=1. Accessed July 29, 2014.
38. "Creation of the Department of Homeland Security." (2012). US Department of Homeland Security. http://www.dhs.gov/creation-department-homeland-security. Accessed July 29, 2014.
39. "Securing the Homeland, Strengthening the Nation." (n.d.) US Department of Homeland Security. http://www.dhs.gov/securing-homeland-strengthening-nation. Accessed July 29, 2014.
40. "Budget-in-Brief." (2014). US Department of Homeland Security. http://www.dhs.gov/sites/default/files/publications/FY15BIB.pdf. Accessed July 29, 2014.
41. "Timeline: A History of the Patriot Act." (n.d). *Al Jazeera America*. http://america.aljazeera.com/watch/shows/the-stream/the-stream-officialblog/2013/10/23/timeline-a-historyofthepatriotact.html. Accessed July 30, 2014.
42. "Did the Sandy Hook Shooting Prove the Need for More Gun Control?" (n.d.). *US News and World Report*. http://www.usnews.com/debate-club/did-the-sandy-hook-shooting-prove-the-need-for-more-gun-control. Accessed July 30, 2014.
43. "Protecting the Force: Lessons from Fort Hood." (2010). US Department of Defense. http://www.defense.gov/pubs/pdfs/dod-protectingtheforce-web_security_hr_13jan10.pdf. Accessed July 31, 2014.
44. "FBI Analysis Finds That Only 4% of Active Shooter Incidents since 2002 Were Perpetrated by Women." (2013). *Public Intelligence*. http://publicintelligence.net/only-4-percent-of-active-shooters-were-women/. Accessed July 31, 2014.
45. Gaffney, James P. (2014). "Preventing the Next Active Shooter Incident—Now Is the Time." *Law Enforcement Today*. http://www.lawenforcementtoday.com/2014/06/07/preventing-the-next-active-shooter-incident-now-is-the-time/. Accessed July 31, 2014.
46. Chakraborty, Barnini. (2013). "White House's $100 Million Mental Health Pledge: Token Gesture or Real Effort at Reform." *Fox News*. http://www.foxnews.com/politics/2013/12/12/token-gesture-white-house-100-million-pledge-for-mental-health-programs. Accessed July 31, 2014.
47. Public Intelligence Report. (2013) https://publicintelligence.net/only-4-percent-of-active-shooters-were-women/. Accessed August 1, 2014.
48. Rich, Sarah. (2014). "Legislating School Safety." *Emergency Management Magazine*, January/February.
49. "High School Facts at a Glance." (2014). US Department of Education. https://www2.ed.gov/about/offices/list/ovae/pi/hs/hsfacts.html. Accessed July 31, 2014.

Section III

Emerging Global Threats

10

Economies of Disasters

Fundamental to our analysis is the assumption that the population... calculates costs and benefits to the extent that they can be related to different courses of action, and makes choices accordingly.... Consequently, influencing popular behavior requires neither sympathy nor mysticism, but rather a better understanding of what costs and benefits the individual or the group is concerned with and how they are calculated.

~David Gladwell, *David and Goliath: Underdogs, Misfits, and the Art of Battling Giants* [1]

ECONOMIC PRINCIPLES

The economy is all around us. It loosely defines the relationship between individual wants, needs, and desires through exchanges of various forms. In other words, the economy is the "production and consumption of goods and the transfer of wealth to produce and obtain those goods" through market interactions "to get what they [the individuals] want or accomplish certain goals" [2]. Economics, or the study of this process, like all social sciences is extremely broad with many facets from broad, worldwide impacts (e.g., gross domestic product, global recessions) to localized or even individual choices (e.g., buying bread at the store). Unfortunately, many emergency management and homeland security professionals have only a rudimentary understanding or consideration of economics before disasters and an even murkier view of how economic processes are impacted (and/or occur) during response and recovery efforts.

This chapter will further consider these issues and project how they will impact risk management in the future.

At the broadest level, there are two major types of economics: macroeconomics and microeconomics. As the names suggest, these two types of economic understanding take a wide look at economic relations (macro) or at the small, localized economic interactions (micro). Specifically, macroeconomics considers the economic activity of an entire region, discipline, country, or international marketplace. This includes how resources are utilized, labor is divided, and leisure and investment occur. On the other hand, microeconomics focuses on individual actions—specifically the dynamics between buyers and sellers [2]. Microeconomics is typically defined by a variety of characteristics including market equilibrium (supply and demand), price elasticity, scarcity of supplies, competition, and game theory [3]. The challenge during disasters is that both microeconomic and macroeconomic dynamics are altered and often need unpredictable or significantly different support to correct, balance, or find equilibrium during recovery.

The most well known of these microeconomic characteristics is the supply versus demand curve. The supply curve represents a tendency of producers to generate or provide an increasing amount of material or resources if the incentive (or pay) increases in a corresponding manner. Conversely, consumers will obtain less of a given resource as the cost or resource input increases. In both cases, these factors are impacted by current and future conditions as well as the quality and preferences of a given community or population. When considering both the supply processes and the demand tendencies, an equilibrium will be found that meets both the consumer and producer needs such that an efficient and effective position can be found [4]. In ideal circumstances, the achievement of market equilibrium is fluid and dynamic; however, it is thoroughly compounded by the impacts of disaster.

The second major consideration of microeconomics is supply-and-demand elasticity. Demand elasticity relates to the sensitivity of the demand of an item to shift when the value or price shifts. Demand is inelastic if it does not respond significantly when values change. Conversely, demand is elastic if it shifts significantly when value changes. Typically, demand is elastic for luxury items, when close alternatives exist, and as people are freer to adjust their behavior [5]. Interestingly, elasticity can shift during disasters. For example, the psychological and sociological definitions of a luxury item can shift significantly during disasters as something that would be routinely and widely available can become difficult to acquire and highly valuable during a disaster. Supply elasticity is similar to demand elasticity, but from the opposite perspective. Specifically, elasticity of supply measures the change in quantity in response to a value or price change. Elasticity of supply is defined by the nature of commodities (perishable or durable goods), time, mobility, number of markets, and scale of production. Several of these have significant impacts after disasters [6]. For example, time is an element of short supply and therefore extremely elastic for supply and demand.

In Other Words...Two Types of Game Theory

In principle, any sequential game that ends after a finite sequence of moves can be "solved" completely. We determine each player's best strategy by looking ahead to every possible outcome. Some games... are...not challenging...[while] other games...are too complex to perform.... Therefore, the players look a few moves ahead and try to evaluate the resulting positions on the basis of experience. In contrast to the linear chain of reasoning for sequential games, a game with simultaneous moves involves a logical circle. Although the players act at the same time, in ignorance of the others' current actions, each must be aware that there are other players who are similarly aware, and so on.... Therefore, each must figuratively put himself in the shoes of all and try to calculate the outcome. His own best action is an integral part of this overall calculation.

~Avinash Dixit and Barry Nalebuff [7]

The last component of microeconomics that will be considered in this chapter is game theory (see Figure 10.1). Game theory attempts to create a science of strategy that helps determine mathematical and logical actions that individuals (aka "players") should take to secure the best outcome in a given situation (aka "game") [7]. Under ideal circumstances, there are three major game scenarios: prisoners' dilemma, mixing moves, and strategic moves. The prisoners' dilemma presents an equal opportunity to two parties or players (aka prisoners) where each will benefit if it can be the first to accomplish the issue. Mixing moves reflects the need for a variety of approaches to minimize the exploitation of a process by an opposing party or rival. Lastly, strategic moves leverage threats or promises to alter

Figure 10.1 The concept of games is an excellent analogy for certain economic concepts and tendencies. (Source: US Air Force.)

expectations of the future, but only work if said processes are credible [7]. However, as has been discussed earlier, these economic processes assume ideal conditions and certainly do not reflect the dynamics of disasters. For example, the incentivization of self-preservation may significantly alter both the prisoners' dilemma and strategic approaches to playing the proverbial economic "game."

In contrast to microeconomics, macroeconomics considers those factors that influence widespread economies at country or global levels. Major factors of macroeconomics include a consideration of gross domestic product (GDP), inflation, deflation, and Keynesian factors, among other things [8]. Loosely stated, a country's GDP is an estimate of the total monetary value of all the finished goods and services that are produced within a 1-year period [9]. Likewise, inflation and deflation represent the community price for a given item relative to its actual value. For example, if inflation is occurring, the price is rising relative to its actual value. Deflation is simply the decrease in price relative to actual value [10]. Lastly, Keynesian factors are also relevant to macroeconomic observations. Keynesian factors consider how government programming and expenditures influence large-scale economies and are often applied during periods of depressed regional or national economies [11].

Each of these will be further evaluated as they relate to disaster response, recovery, and mitigation.

There has been significant study of and research into how disasters impact GDP. For example, according to a study by the Brookings Institution, the quantitative impact of disasters on large and diversified economies (e.g., the United States) is relatively minor. Even if the ability to produce and consume goods is impacted in a large area, it still remains only a very small fraction of an overall national or global economy. Even significant physical and cultural disasters like the 9/11 terrorist attacks only impacted GDP by 0.5% [12]. However, there have been noted events such as the earthquake and tsunami in Japan that impacted GDP as much as 0.7% to 3.0%. Likewise, even though New Orleans has had some nominal GDP recovery from Hurricane Katrina, the population is still 30% less than it was before the disaster [13].

Likewise, most economists agree that impacts to inflation and deflation from disasters are only temporary. These temporary inflationary pressures typically relate to the rising costs of food products, agriculture, and other consumables necessary for basic functions such as food, water, and shelter. This is especially true as the demand for construction goods and labor increases as homes, businesses, schools, and other community buildings are rebuilt [14]. For example, during a cyclone in Fiji, the annual food prices were driven up by nearly 7%; however, the effects were quickly neutralized with time and external economic forces [15]. Disaster-related inflation can also occur from an increased debt burden to a given country from the massive incident-related losses and related costs of recovery [16]. However, this particular risk is often mitigated through government actions or international support and aid, which will be discussed next.

KEYNESIAN ECONOMICS

The Keynesian economic model was developed during the 1930s by British economist John Maynard Keynes as an attempt to understand the role of government in the Great Depression. Specifically, Keynes advocated for increased government spending and lower taxes to stimulate demand and ultimately positively impact the global economy (see Figure 10.2). However, since that time, the term "Keynesian economics" has been used to refer to the "concept that optimal economic performance could be achieved…by influencing aggregate demand

Figure 10.2 Maynard Keynes advocated for increased government spending and lowering taxes to stimulate demand and ultimately positively impact the global economy.

through activist stabilization and economic intervention policies by the government" [11]. Over much of the last century this initial model has been modified and altered to address broader economic dynamics than initially considered during the Great Depression.

To understand Keynesian models, there must be a foundational understanding of free-market economies that generate adequate resources to provide financial stability and a strong workforce. Consequently, if there is high unemployment in this type of environment, it is due to a lack of economic incentives to drive private employers to increase their activities and thus hire more employees. Therefore, Keynesian models assume that the only way to shift this process is to "increase aggregate demand by increasing expenditures in the economy" [17]. Another essential element to most modern applications of Keynesian economic models is the idea of a paradox of thrift. Specifically, a paradox of thrift is the concept that there is an increasing desire to minimize spending (both personally and organizationally) and save funds while the strength of the economy is depressed, which in turn contributes to the decline of GDP and other macroeconomic indicators. However, the model predicts that with government injections and interferences into the natural economic market, the health of the economy will recover more quickly and ultimately lead to a multiplier effect, leading to a bigger final economy [18].

While Keynesian economic models are not necessarily a reflection on the size of government (the base principles can be applied to both lean and big governments), there are two elements that are easily understood about how big government applies these principles. First, government spending and legislative (or executive) programming can potentially reduce unemployment by creating jobs. Secondly, when government is big and centralized it is fully capable of using a financial "fire hose" to fund programs and services to supplement the economy and other related efforts [19]. The most recent example of this approach occurred in 2008, when US President Barack Obama urged Congress to pass an $800 billion stimulus package that would support thousands of "shovel-ready" jobs where people could (theoretically) be put right to work [20] (see Figure 10.3). Likewise, regardless of the method of determination, the size of the US government (and many developing countries throughout the world) has increased significantly with rare leveling after Keynesian bumps in spending and programming.

However, over the last decade, an increasing number of groups (usually politically conservative) have objected to the government-spending Keynesian models of economic recovery. Specifically, arguments against Keynesian models are based around the concept of where the funds that ultimately stimulate the economy originate from. For example, government funds are available through the taxation of individuals and business corporations. Therefore, the funds that are used to stimulate the economy are ultimately simply redistributed within

Figure 10.3 President Obama (Pictured here with FEMA Administrator Craig Fugate) exercised a Keynesian economic approach in 2008 by approving an $800 billion "shovel ready" jobs program. (Source: White House/Pete Souza.)

a given economy. While Keynesian dissenters certainly acknowledge the right of government to leverage taxation revenue to the benefit of the economy, they strongly object to the ultimate effectiveness of this approach. For example, the previously discussed $800 billion Keynesian-based stimulus pushed by President Obama and passed in 2008 was intended to keep unemployment from exceeding 8%; however, unemployment rose above 10% by 2010 [21].

In Other Words...Why Government Spending Does Not Support Economic Recovery

Moving forward, the important question is why government spending fails to end recessions. Spending-stimulus advocates claim that Congress can "inject" new money into the economy, increasing demand and therefore production. This raises the obvious question: From where does the government acquire the money it pumps into the economy? Congress does not have a vault of money waiting to be distributed. Every dollar Congress injects into the economy must first be taxed or borrowed out of the economy. No new spending power is created. It is merely redistributed from one group of people to another.

~Brian M. Riedl [21]

Regardless of the impact or appropriateness of Keynesian approaches to the economy, government provides a vital role in the economic processes that occur after a disaster occurs. In 2011, the US Office of Management and Budget reported to Congress about the spending on disaster relief over the previous decade. Their report found that the average annual expenditure was approximately $11.5 billion. However, this estimate did not include catastrophic events like Hurricanes Katrina and Rita. Likewise, the Center for American Progress found that the US federal government spent $136 billion from 2011 to 2013, which is approximately $400 per American household per year. These funds were distributed through 96 different programs supported by 19 different departments or agencies, including the Departments of Agriculture, Homeland Security, and Labor, and the Social Security Administration, to name a few [22]. Interestingly, over that evaluation period, 11 years required supplemental funding in addition to what was originally appropriated [23].

PUBLIC–PRIVATE PARTNERSHIPS IN RECOVERY

While it is unlikely that government financial support for impacted individuals and communities will stop, there is an increased opportunity for third-party participation in the disaster relief process. For example, several states, including Alabama and Virginia, have established public–private disaster relief funds to be used as a so-called safety net for those individuals who are impacted by disasters but are ineligible for federal disaster recovery funds for one reason or another. For example, former Virginia governor Bob McDonnell created the Virginia Disaster Relief Fund in 2011 after the state was denied federal assistance (see Figure 10.4). This fund was initially supported through state government budgetary surpluses, funds collected at state-licensed liquor stores, and small and large private donations [24]. Likewise, famous entertainers routinely establish disaster relief concerts to raise funds and awareness for the communities impacted. For example, a disaster relief concert was established in late 2012 to raise funds for those impacted by Superstorm Sandy. As a result, more than $73 million was distributed to over 400 different relief organizations that ultimately helped 54,000 families that had returned to impacted homes and provided medical services to more than 65,000 individuals [25].

Figure 10.4 After former Virginia governor Bob McDonnell was denied federal disaster assistance for a local disaster, he created the Virginia Disaster Relief Fund to allow for nontraditional public and private donations. (Source: US Navy/ Mass Communication Specialist 2nd Class Michael R. Hinchcliffe.)

Likewise, numerous state and regional organizations have been created around the idea of private organizations in loose affiliations to prepare for disasters and protect critical infrastructure. For example, a group called ChicagoFIRST was started in 2003 to focus on financial institutions in Chicago and identify mechanisms for these institutions to create strategic partnerships among themselves and with public preparedness and prevention agencies like the US Department of Homeland Security and the US Treasury Department [26]. Similar groups like the FBI's InfraGard program have created strategic partnerships to exchange information and focus on emerging threats and issues like cyberterrorism. In the case of InfraGard there are 85 chapters with more than 35,000 members that are closely connected regionally and loosely affiliated nationally [27].

Another way to facilitate public–private activities for disaster response and recovery is the use of fund-raisers such as telethons and text-to-donate campaigns. Without fail, famous musicians, actors, and artists typically leverage their fame to drawn attention to the disaster-related cause. Likewise, nongovernmental organizations often partner with these groups or private companies to create synergy around the desire of the general public to aid those in need. For example, after Superstorm Sandy impacted much of the northeast United States in 2012, the American Red Cross received more than $300 million for relief and recovery efforts generated from telethons, with a majority of the funds going to address food, shelter, and other types of individual assistance [25].

From a slightly different perspective, the public–private economy also comes into play when nonprofit organizations and/or nongovernmental organizations (NGOs) solicit donations from citizens and businesses alike to utilize in disaster-impacted areas. Traditionally, these funds have been defined, managed, and controlled by traditional economic forces. However, with the rise of mobile phones and other digital technologies, groups like the American Red Cross have shifted to allow a new donations economy built around these technologies. For example, the American Red Cross set up a text-to-donate campaign for the 2010 Haiti earthquake that caused significant infrastructure damage and loss of life. As a result, nearly $32 million was generated within the first month [28]. In the time since that event, text-to-donate campaigns have become common practice for nearly all major NGOs, including UNICEF, the UN World Food Programme, Save the Children, and the Salvation Army, to name a few [29].

However, these financial interactions during and after disasters are not considered wholly efficient. Several different studies conducted by social scientists and economists have questioned how government support creates disincentives for individuals and private support to better prepare financial preparedness. For example, some of those surveys have suggested that when postdisaster government relief is expected, individuals (or related families) save significantly less money for "rainy day" funds and underinvest in insurance to cover potential losses. One empirical study indicated that a $1,000 increase in the Federal Emergency Management Agency (FEMA) individual assistance (IA) grants resulted in a $6,350 decrease in average insurance coverage [30]. These dynamics are important moving into the future as the government seeks methodologies to support impacted communities, but via shared and efficient processes.

SHARED, VIRTUAL, AND EXPERIENCE ECONOMY MODELS

With the rise in technology and the availability of information exchange via mobile devices, the traditional microeconomic and macroeconomic models have been altered. The incentives and disincentives that drive those decisions are significantly changed via individual empowerment. No longer are individuals limited to having their resources, employment, purchases, and financial management fully controlled by large (and often disembodied) economic forces such as government or large investors. As will be discussed in this section, many of the definitions of currency and markets have been redefined to be much broader, inclusive, and, ultimately, preindustrial.

In his book *Who Owns the Future?* Jaron Lanier presents the juxtaposition of the Kodak Company and Instagram to address this revolutionary change. With the establishment of the Eastman Kodak company in 1892, a technological innovator was started with some of the earliest and most effective photography film (see Figure 10.5). For much of the next century, Kodak was the leading provider of film products and became synonymous with success and quality [31]. Kodak eventually invented the digital camera, which began a revolution in photography. At its height, Kodak employed more than 140,000 individuals and was worth more than $28 billion [32]. However, by 2013 Kodak had filed for bankruptcy and had reduced its workforce to 4,700, which was roughly equal to its size in 1908 [31,33]. To contrast this rise and fall, Lanier considered the rise in technology and availability of high-quality, easy-to-use, ubiquitously available cameras built into cell phones that were widely available

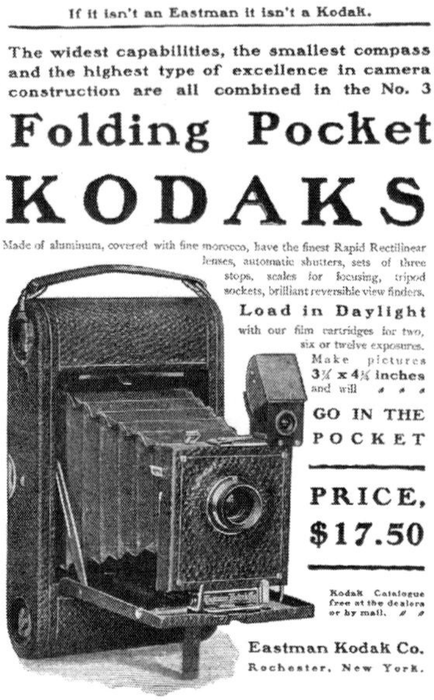

Figure 10.5 After its establishment in 1892, the Eastman Kodak company became a leading provider of film products for much of the twentieth century before it was replaced by quicker and more pervasive technologies.

throughout the world. These devices allow for access to photography applications like Instagram. In the case of Instagram, it was sold to Facebook in 2012 (just prior to the Kodak bankruptcy filing) for $1 billion with a mere 13 employees supporting the services and functions of the app [32]. No longer are high-production, industrial-driven economic models necessary for success. The availability and access to services and information are just as valuable.

As clearly seen in the example of Instagram, the concept of production has changed. In a traditional manufacturing process, the manufacturer invests financially to develop the design, production, and shipping of individual physical units, which are then sold individually or in bulk within the broader economy [34]. The quintessential example of this concept is assembly line processes where individual parts are manufactured and then put together in a sequential fashion and sold as one cohesive piece. In contrast, the digital world has created a wholly different production process to run parallel to the traditional process. In a digital world the production is virtual and tied together through computer coding of binary ones and zeros, with singular cohesive pieces far less common. Moreover, a digital production process inherently requires a high level of creativity and design, and the user (or consumer) interface of this production changes regularly even though the production or exchange of information does not. Most importantly, the digital production process requires significantly less financial or resource input. In contrast to the traditional manufacturing production, the cost of digital production is effectively zero, requiring only the time, energy, and passion of the developer(s).

In Other Words...Digital Production

We define digital production as the process by which creative ideas and assets (images, text and interactive apps) are translated into an array of digital media—web sites...email, mobile and social media applications—so that the right messages are delivered via the right channel at the right time to the right users.

~Neal Prescott [35]

It is this shift to digital or virtual production that has created so many new variants of economic exchange and currency. For example, forms of digital currency (officially known as cryptocurrency) like bitcoin have emerged over the last several years (see Figure 10.6).

Figure 10.6 Digital currency like the bitcoin has been developed as mobile and wireless technologies have increased in usage and availability.

These currency systems are unregulated by banks or government bodies and are issued and usually controlled by digital producers or developers. For example, bitcoin allows users to mine, buy, sell, and accept bitcoin currency in exchange for a virtual signature that validates and tracks its movement [36]. Unfortunately, the potential benefits of virtual currency like bitcoin have not yet been fully worked out. They have thus far been difficult to obtain (typically through middlemen only) and subject to wide swings in price and valuation. Moreover, it is not widely accepted whether virtual currency will serve as the basis for future economies or merely serve as a "hyperefficient online transaction system" [37]. Even the legality is not universal. While the United States, Germany, Finland, Singapore, and Canada have issued tax guidance or legal clarification on its use and Ireland, Israel, and Slovenia are considering it, the rest of the world has not yet clarified how digital currency like bitcoin can be deployed within current economic markets [38].

In Other Words...Characteristics of Bitcoins

> A bitcoin...has no intrinsic value...is not regulated or backed by any real-world commodity...or any central bank or government or pegged to any real-world currency...[and yet] it has the characteristics of money...[and] is based on a decentralized, peer-to-peer network.
>
> ~David Richardson and Alesya Tepikina [36]

The challenge for the bitcoin is how it will be adapted in the future and what the risk or impact is to communities and therefore what emergency management and homeland security professionals should know. For example, only 76% of surveyed Americans were familiar with bitcoin; however, there was a significant spike in 2013–2014 that drove the price of bitcoins from $13 per unit to as high as $1,200 per unit. While there are only 63,000 bitcoin transactions per day internationally (which is significantly less than the millions conducted by credit card companies), bitcoin is already highly competitive in emerging financial markets that represent more than one-third of the world's population. Unfortunately, bitcoin does have a growing reputation as the preferred methodology for online illegal activities such as the buying and selling of illegal drugs and weapons. Moreover, because of the unregulated processes there is a higher chance of fraudulent exchanges through middleman involvement [39].

These present significant risks that must be considered. However, the benefit to emerging economies and disaster-impacted areas may ultimately significantly impact its use, effectiveness, and long-term longevity.

In addition to the establishment of virtual currencies like bitcoin, another "new" economy has also developed based around the idea of shared resources and currencies. This so-called "gift economy" is not based on traditional models of currency usage to purchase or acquire raw or manufactured goods. One of the most profound examples of this model was developed by the food retail company Panera Bread. Specifically, three "Panera Cares Cafes" were opened in St. Louis, Missouri; Detroit, Michigan; and Portland, Oregon, with a "Pay What You Can" transaction model. According to Panera, 60% of people paid the suggested price, 20% paid less, and 20% paid more, with an overall income of slightly more than the national storage average [41]. At its surface, a gift economy is a form of generosity or charitable giving, but that does not fully incorporate the gift economy, as other services of value can also be provided in gift form. For example, other models of a gift economy include open-source coding and development as well as sites like WikiLeaks as one individual contributes a thing of value for no direct currency exchange [40].

In Other Words...Definition of Economy

The new exchange systems we are exploring blur the boundary between the monetary and non-monetary realms and therefore the standard definition of the "economy." Really, what is the economy? Underneath the ephemera of money—slips of paper, bits in computers—what changes when the economy grows or shrinks? Ultimately, what economics attempts to measure, underneath money, is the totality of all that human beings make and do for each other.

~Charles Eisenstein [41]

A closely related alternative economy is the so-called barter or peer-to-peer economy, which is the exchange of one valued item for another. While often associated with precurrency cultures (e.g., Pilgrims and Native Americans), the concept has grown significantly in its use, particularly through the integration of technology. Specifically, anything and everything can be rented, shared, sold, or exchanged via the Internet. One of the most well-known conduits for this behavior is a site called Airbnb. The Airbnb service allows people to post homes, rooms, or residences they own to be rented directly to another individual. By the end of 2013, more than 10 million people had stayed in one of Airbnb's 550,000 listed homes in more than 175 countries [42]. Similar services like Uber, Lyft, and Sidecar attempt to allow people to rent others' vehicles for various purposes (with or without a driver). While it is not totally clear how many people leverage these services for rides, the availability of the services is growing with availability in more than 60 cities in 25 different states [43]. Moreover, one study by the Knight Foundation found that this type of peer-to-peer transaction increased 36% from 2009 to 2012 with no signs of slowing down [44]. This type of sharing is nearly endless, with additional services already available for pet sitting, task assignments, and high-end wardrobes.

The shareable economy is also of particular importance to the emergency management community. For example, Airbnb has memorandums of understanding in communities like Portland, Oregon, and San Francisco, California, to provide assistance during disaster response or recovery. Specifically, Airbnb agreed to identify potential residential hosts for emergency workers and survivors, provide emergency preparedness materials to host sites, and alert hosts and guests about localized hazards [45]. Specifically, Airbnb deploys a "disaster response mode' that is deployable within 30 minutes and provides a single landing page for those seeking assistance and waives all fees associated with the rental and transaction. These Airbnb agreements grew out of their response to Superstorm Sandy, where their users helped facilitate the sheltering of 1,400 displaced persons [46]. This type of approach could be highly impactful to emergency managers considering methods not only to shelter impacted community members, but also to evacuate areas of risk in efficient and effective ways.

CROWDFUNDING

The last components of modern and emerging technologies that will be considered are the concepts of crowdfunding and digital bridges. Much like crowdsourcing, which has been discussed off and on throughout this book, crowdfunding uses the collective funding power of a group of interested people to commit funds or financial means to projects, concepts, or other initiatives that otherwise might be deemed inappropriate or inefficient to be funded through traditional supply and demand models.

The most well-known crowdfunding system is Kickstarter, which generated more than $480 million from three million people in 2013. This figure works out to $1.3 million per day or $913 per minute, with nearly 20,000 projects receiving full funding. While there have been only a handful of truly notable projects, the funding has increased by more than 50%, with increasingly unique and specific projects being presented for consideration everyday [47].

Interestingly, crowdfunding has already been utilized in disaster response and recovery. For example, after Superstorm Sandy in 2012, hundreds of disaster victims initiated personalized disaster relief funding campaigns on sites such as GoFundMe, IndieGoGo, and HelpersUnite, which ultimately yielded a few million total donated dollars. While this total is not significant in the grand spectrum of a large-scale disaster, it does show proof of the concept in the possibility of crowdfunding for disaster relief [48]. Likewise, other groups utilized crowdfunding to provide additional fund-raising for response agencies like the American Red Cross [49]. Similar emergent postdisaster crowdfunding efforts have arisen over the last several years for disasters of various sizes in geographic areas throughout the world. Likewise, other crowdfunding efforts have arisen around the need for more cost-effective sheltering and energy generation in disaster-impacted areas [50].

The question for emergency managers is how to handle crowdfunding. To date, crowdfunding of government-based projects has been limited in application and scope. For example, between 2010 and 2014, there were more than 1,000 civic projects funded for a total of nearly $11 million (or $6,357 per project) [51]. While these figures might seem impressive, they are only a small fraction of those projects funded for private or commercial activities. Moreover, while communities like San Francisco have been much more aggressive in their approach to crowdfunding, it is unclear how quickly smaller and more homogenous communities will adopt these approaches. Additionally, it leaves open the question of how to address public organizations or individuals who leverage crowdfunding after a disaster. Do emergency management organizations embrace these approaches as effective community preparedness, ignore them as meaningless activities in the overall recovery effort, or publicly advise against such actions due to concerns about legality, ethics, or simply distractions?

Understanding these economic considerations as well as all those discussed in this chapter is a critical step in the future of emergency management. Understanding and ultimately forecasting trends in these areas will help in all phases of disaster management. Too often, emergency managers and the communities they prepare for are ill prepared to understand and respond to the economic changes that occur. Like many of the characteristics needed to be well-rounded emergency managers and homeland security officials, economics—global, local, and individualized exchanges—needs to be studied and observed with greater focus.

REFERENCES

1. Gladwell, David. (2013). *David and Goliath: Underdogs, Misfits, and the Art of Battling Giants*. New York: Little, Brown and Company.
2. "What Is Economics? A Definition of Economics." (n.d.). http://www.whatiseconomics.org. Accessed August 1, 2014.
3. "Microeconomics." (2014). Khan Academy. https://www.khanacademy.org/economics-finance-domain/microeconomics. Accessed August 1, 2014.
4. "Supply, Demand, and Market Equilibrium." (2014). Khan Academy. https://www.khanacademy.org/economics-finance-domain/microeconomics/supply-demand-equilibrium. Accessed August 1, 2014.
5. "Elasticity." (n.d.). Ohio State University. http://www.econ.ohio-state.edu/jpeck/H200/EconH200L5.pdf. Accessed August 3, 2014.
6. "Elasticity of Supply: Meaning, Types, Measurement, Determinants and Significance." (n.d.). Sundaramponnusamy Blog. http://sundaramponnusamy.hubpages.com/hub/Elasticity-of-Supply-Meaning-Types-Measurement-Determinants-and-Significance. Accessed August 3, 2014.
7. Dixit, Avinash and Nalebuff, Barry. (2008). "Game Theory." *The Concise Encyclopedia of Economics*. http://www.econlib.org/library/Enc/GameTheory.html. Accessed August 3, 2014.
8. "Macroeconomics." (2014). Khan Academy. https://www.khanacademy.org/economics-finance-domain/macroeconomics. Accessed August 4, 2014.
9. "A Glossary of Political Economic Terms: Gross Domestic Product." (2005). Auburn University. http://www.auburn.edu/~johnspm/gloss/GDP. Accessed August 4, 2014.
10. "Inflation—Measuring the Cost of Living." (2014). Khan Academy. https://www.khanacademy.org/economics-finance-domain/macroeconomics/inflation-topic. Accessed August 4, 2014.
11. "Keynesian Economics." (2014). Investopedia. http://www.investopedia.com/terms/k/keynesianeconomics.asp. Accessed August 4, 2014.
12. Roberts, Bryan W. (2009). "The Macroeconomic Impacts of the 9/11 Attack: Evidence from Real-Time Forecasting." https://www.dhs.gov/xlibrary/assets/statistics/publications/ois_wp_impacts_911.pdf. Accessed August 5, 2014.
13. Baily, Martin Neil. (2011). "Can Natural Disasters Help Stimulate the Economy?" Brookings Institution. http://www.brookings.edu/research/opinions/2011/09/01-disasters-stimulus-baily. Accessed August 4, 2014.

14. Parsons, Stephanie. (2011). "RBA Economics Competition 2011." Griffith University. http://www.rba.gov.au/econ-compet/2011/pdf/second-prize.pdf. Accessed August 5, 2014.
15. Pelling, Mark, Ozerdem, Alpaslan, and Barakat, Sultan. (2002). "The Macro-Economic Impact of Disasters." Progress in Development Studies. https://www.google.com/url?sa=t&rct=j&q=&esrc=s&source=web&cd=1&cad=rja&uact=8&ved=0CB0QFjAA&url=http%3A%2F%2Fwww.researchgate.net%2Fpublication%2F235737565_The_Macro-economic_impact_of_disasters%2Flinks%2F00b7d52c9de15b02b8000000&ei=DgbhU_SWAZHgsATo7oKwBQ&usg=AFQjCNHiFRpH73ZBhg9F-prNCCV0nwMrZw&bvm=bv.72197243,d.aWw. Accessed August 5, 2014.
16. Popp, Aaron. (2006). "The Effects of Natural Disasters on Long Run Growth." http://business.uni.edu/economics/themes/popp.pdf. Accessed August 5, 2014.
17. Nanda, Sushil. (2012). "10 Most Essential Features of Keynesian Analytical Framework." Preserve Articles. http://www.preservearticles.com/201102083969/10-most-essential-features-of-keynesian-analytical-framework.html. Accessed August 5, 2014.
18. Pettinger, Tejvan. (2008). "Keynesianism vs. Monetarism." Economic Help. http://www.economicshelp.org/blog/1113/concepts/keynesianism-vs-monetarism. Accessed August 6, 2014.
19. Aziz, John. (2014). "Why Do Keynesians Prefer Government Spending over Private Spending?" *The Week*. http://theweek.com/article/index/256972/why-do-keynesians-prefer-government-spending-over-private-spending. Accessed August 6, 2014.
20. Worstall, Tim. (2013). "The Reason That Shovel Ready Stimulus Didn't Work Is That There Wasn't Any Stimulus." Forbes. http://www.forbes.com/sites/timworstall/2013/11/01/the-reason-that-shovel-ready-stimulus-didnt-work-is-that-there-wasnt-any-stimulus. Accessed August 6, 2014.
21. Riedl, Brian M. (2010). "Why Government Spending Does Not Stimulate Economic Growth: Answer the Critics." The Heritage Foundation. http://www.heritage.org/research/reports/2010/01/why-government-spending-does-not-stimulate-economic-growth-answering-the-critics. Accessed August 6, 2014.
22. Weiss, Daniel J. and Weidman, Jackie. (2013). "Disastrous Spending: Federal Disaster-Relief Expenditures Rise amid More Extreme Weather." Center for American Progress. http://www.americanprogress.org/issues/green/report/2013/04/29/61633/disastrous-spending-federal-disaster-relief-expenditures-rise-amid-more-extreme-weather. Accessed August 7, 2014.
23. Lindsay, Bruce R. (2014). "FEMA's Disaster Relief Fund: Overview and Selected Issues." Congressional Research Service. http://fas.org/sgp/crs/homesec/R43537.pdf. Accessed July 7, 2014.
24. "Virginia Establishes Public–Private Disaster Relief Fund." (2011). *Philanthropy News Digest*. http://www.philanthropynewsdigest.org/news/virginia-establishes-public-private-disaster-relief-fund. Accessed August 7, 2014.

25. McKay, Hollie. (2013). "Celebs Raised Millions for Sandy Relief So Where Did It Go?" *FoxNews*. http://www.foxnews.com/entertainment/2013/10/29/celebs-raised-millions-fo-sandy-relief-so-where-did-it-go. Accessed August 7, 2014.
26. "History of ChicagoFIRST." (2014). ChicagoFIRST. https://www.chicagofirst.org/about/about_us.jsp. Accessed August 12, 2014.
27. "InfraGard: A Partnership That Works." (2010). Federal Bureau of Investigation (FBI). http://www.fbi.gov/news/stories/2010/march/infragard_030810. Accessed August 12, 2014.
28. Crowe, Adam. (2012). *Disasters 2.0: The Application of Social Media in Modern Emergency Management*. Boca Raton, FL: CRC Press.
29. Lackey, Katharine. (2013). "How You Can Help Typhoon Haiyan Survivors." *USA Today*. http://www.usatoday.com/story/news/world/2013/11/09/haiyan-how-you-can-help/3484467. Accessed August 12, 2014.
30. Kousky, Carolyn, Michel-Kerjan, Erwann O., and Raschky, Paul A. (2013). "Does Federal Disaster Assistance Crowd-Out Private Demand for Insurance?" Wharton School University of Pennsylvania. https://www.google.com/url?sa=t&rct=j&q=&esrc=s&source=web&cd=10&cad=rja&uact=8&ved=0CIABEBYwCQ&url=https%3A%2F%2Fwww.aeaweb.org%2Faea%2F2014conference%2Fprogram%2Fretrieve.php%3Fpdfid%3D102&ei=Ya_jU5yZAey_sQS_tIAQ&usg=AFQjCNGTuGcOesei8x2iiTR-iy6p98Eu4A&bvm=bv.72676100,d.aWw. Accessed August 12, 2014.
31. "1878–1929." (n.d.). Kodak. http://www.kodak.com/ek/US/en/Our_Company/History_of_Kodak/Milestones_-_chronology/1878-1929.htm. Accessed August 12, 2014.
32. Lanier, Jaron. (2013). *Who Owns the Future?* New York: Simon & Schuster.
33. Daneman, Matthew. (2013). "Kodak Bankruptcy Officially Ends." *USA Today*. http://www.usatoday.com/story/money/business/2013/09/03/kodak-bankruptcy-ends/2759965/. Accessed August 12, 2014.
34. Drexler, K. Eric. (2013). *Radical Abundance: How a Revolution in Nanotechnology Will Change Civilization*. New York: Public Affairs.
35. Prescott, Neal. (2011). "What Is Digital Production?" Deliver Blog. http://blog.deliveroffshoring.com/?p=126. Accessed August 14, 2014.
36. Richardson, David and Tepikina, Alesya. (2013). "Bitcoin—Virtual Currency of the Future? Part One." Dorsey. http://www.dorsey.com/eu_cm_bitcoin_virtual_currency. Accessed August 14, 2014.
37. McMillan, Robert. (2014). "The Fierce Battle for the Soul of Bitcoin." *Wired*. http://www.wired.com/2014/03/what-is-bitcoin/?mbid=social_fb. Accessed August 14, 2014.
38. Hill, Kashmir. (2014). "Bitcoin's Legality around the World." *Forbes*. http://www.forbes.com/sites/kashmirhill/2014/01/31/bitcoins-legality-around-the-world/. Accessed August 14, 2014.
39. Boring, Perianne. (2013). "Bitcoin Basics for the '76 Percenters' Who Don't Have a Clue What It Is." *Forbes*. http://www.forbes.com/sites/perianneboring/2014/02/22/bitcoin-basics-for-the-76-percenters-who-dont-have-a-clue-what-it-is. Accessed August 14, 2014.

40. "Panera Cares Lets Customers Set Prices." (2013). Here and Now. http://hereandnow.wbur.org/2013/03/26/panera-cares-cafe. Accessed August 14, 2014.
41. Eisenstein, Charles. (2011). "Sacred Economics: Chapter 16, Transition to Gift Economy." Sacred Economics. http://sacred-economics.com/sacred-economics-chapter-16-transition-to-gift-economy/. Accessed August 14, 2014.
42. Grant, Rebecca. (2013). "To Infinity and Beyond: Airbnb Books 10M Guests in Its Castles, Cabins, and Couches." Venture Beat. http://venturebeat.com/2013/12/19/to-infinity-and-beyond-airbnb-books-10m-guests-in-its-castles-cabins-and-couches/. Accessed August 15, 2014.
43. "Cities We're In." (2014). Lyft. https://www.lyft.com/cities. Accessed August 15, 2014.
44. Shueh, Jason. (2014). "Why Everyone Is Watching the Sharing Economy." *Government Technology*. http://www.govtech.com/local/Why-Everyone-is-Watching-the-Sharing-Economy.html. Accessed August 15, 2014.
45. Pittman, Elaine. (2014). "White House Innovation Day Highlights Disaster Response, Recovery." *Emergency Management Magazine*. http://www.emergencymgmt.com/disaster/White-House-Innovation-Day-Disaster-Response.html. Accessed August 15, 2014.
46. Smith, Rory. (2013). "San Francisco's Mayor Lee Launches Sharing Economy Partnership for Disaster Response." Shareable. http://www.shareable.net/blog/san-franciscos-mayor-lee-launches-sharing-economy-partnership-for-disaster-response. Accessed August 15, 2014.
47. Wasserman, Todd. (2014). "Kickstarter Pledges Hit $913 per Minute in 2013, with Few Breakouts." *Mashable*. http://mashable.com/2014/01/08/kickstarter-2013/?utm_cid=mash-com-Tw-main-link. Accessed August 15, 2014.
48. Zongker, Brett. (2012). "Hurricane Sandy Relief Gets Boost from Crowdfunding Efforts." *Huffington Post*. http://www.huffingtonpost.com/2012/12/28/crowdfunding-sandy-relief_n_2376704.html. Accessed August 15, 2012.
49. Luzar, Charles. (2012). "Crowdfunding Will Streamline Aid for Victims of Sandy and Future Disasters." *Venture Beat*. http://venturebeat.com/2012/11/05/crowdfunding-disaster-donations/. Accessed August 15, 2014.
50. Markham, Derek. (2014). "This Start-Up Is Crowdfunding a Smarter Portable Disaster Relief Shelter." *Ecopreneurist*. http://ecopreneurist.com/2014/03/11/startup-building-better-portable-disaster-relief-shelter/. Accessed August 15, 2014.
51. O'Brien, Kathleen. (2014). "Civic Crowdfunding: The Future of Paying for Public Projects." *Nation Swell*. http://nationswell.com/civic-crowdfunding-community-projects-citizens-governments. Accessed August 16, 2014.

11

Sustainability and Environmental Factors

Climate change ranks among the world's most serious problems—such as disease outbreaks, poverty, terrorism and the proliferation of weapons of mass destruction…[and represents] the greatest challenge of our generation.

~John Kerry [1]

The polar bear is us.

~Patricia Romero Lankao [2]

SUSTAINABILITY

The concept of sustainability has gained significant public and media attention over the last decade. Real and targeted attempts to create sustainable processes and organizations have arisen in nearly all communities and industrial sectors. Both private and civic leaders have viewed sustainability for both economic and philosophical reasons as the issue has moved from fringe concept to practical application. Likewise, the concept of sustainability has become increasingly politicized in many communities and public sectors. It is also, as will be discussed in this chapter, inherently tied to environmental risks and the impact of human choices, which have driven the concept of sustainability directly into the future choices of the emergency management community.

To understand these impacts, a clear definition of sustainability must be established. Unfortunately, this is not as easy as it seems. Perhaps the most well-established concept of sustainability was first presented by Dr. Dennis Mileti. Dr. Mileti presented, among other things, a need to create hazard mitigation strategies that are sustainable in each individual community. These strategies were ultimately identified as cross-linked relationships between natural resources (and their risks), local economies, and social resiliency [3]. In short, Mileti specifically created an accurate layering to approaching sustainability that would become inefficient due to lack of support from social, cultural, or economic characteristics within a given community.

From a more practical standpoint, Mileti described six community objectives that must occur simultaneously to mitigate hazards in the United States in a sustainable way. The first of these objectives was to maintain and enhance environmental quality. The second objective for sustainability was to shift human activities in a way that does not reduce the carrying capacity of the ecosystem. If this objective is not met there is the possibility that the impact of the hazards will last longer when present in a given community. The third characteristic presented by Mileti was to maintain and enhance the quality of life in a given area as these factors (such as income, education, health care, housing, and employment) can significantly impact the scope and severity of given hazards. The fourth characteristic of community sustainability was to foster local resiliency and responsibility such that the given community could withstand given hazards with a tolerable level of losses and impact. The fifth characteristic presented by Mileti was to foster local economies that are vibrant and diversified such that postdisaster recovery is more efficient and effective. The final characteristic of sustainability was to ensure that intergenerational equity would ensure hazards are reduced across all ethnic, racial, and income groups and that recovery efforts are not shifted to certain generational or sociodemographic groups [3]. These characteristics create the foundations for sustainability; however, like all aspects, these characteristics are more easily said than done.

In Other Words...Sustainability Defined

Sustainability means that a locality can tolerate—and overcome—damage, diminished productivity, and reduced quality of life from an extreme event without significant outside assistance.

~Dennis Mileti [4]

While Mileti's view of sustainability is the most well defined and understood, there are numerous other ways to view how emergency managers should consider sustainability. One perspective maintains a direct connection between sustainability and resources: specifically, a strong need to avoid the depletion or permanent damage to any natural, human, or technological resource that will be needed to reduce risk or facilitate recovery after a disaster. Moreover, as overly simplistic as it sounds, a critical element of sustainability is the avoidance of hazards, threats, and other elements which create risk. The survival and ultimate sustainability of a community is fully achievable when risk is avoided. In other words, sustainability refers to the "capability of complex systems... to cope with changing conditions, to permanently adapt and, nevertheless, satisfy present needs" [5].

Sustainability is not a static process due to its social, cultural, and economic complexity within a given community. Decision making by politicians, community leaders, and ultimately emergency managers must incorporate sustainability at every step that drives communities in a particular direction. This means that every choice before, during, and after disasters has potential sustainable components and considerations. Sustainable decisions ultimately improve quality of life and the long-term success of a given community, which is beneficial regardless of whether disasters occur or not. As such, many emergency managers have connected sustainable decision making (even though they may not call it that) with whole-community planning and preparedness concepts.

In Other Words...FEMA's Whole-Community Concept

We fully recognize that a government-centric approach to emergency management is not enough to meet the challenges posed by a catastrophic incident. Whole community is an approach to emergency management that reinforces...that we must leverage all of the resources...to prepare for, protect against, respond to, recover from and mitigate against all hazards.... Both the composition of the community and the individual needs of community members, regardless of age, economics, or accessibility requirements, must be accounted for when planning and implementing disaster strategies.

~FEMA [6]

Under the leadership of Administrator Craig Fugate, the US Federal Emergency Management Agency (FEMA) proposed the whole-community concept and began implementation in 2010. This concept arose after major events like Hurricane Katrina that quickly identified the lack of preparedness for various community sectors such as lower socioeconomic sectors and certain racial and cultural subcommunities. As such, there was a significant dedication to identifying a methodology that would embrace the entire community and ultimately ensure a more resilient response and recovery. Much like the sustainability issues already discussed, the whole-community concept brings together disparate parts or gaps in preparedness to ultimately improve the overall readiness (see Figure 11.1).

FEMA's whole-community concept was founded on six principles. The first principle was to understand community complexity. This included the geographic, cultural, language, and socioeconomic factors that create the unique features of a given community or area. Secondly, the whole-community concept required the recognition of community capabilities and needs. This included government and private resources as well as those features and characteristics that routinely define the community and

Figure 11.1 Under the leadership of Administrator Craig Fugate, the US Federal Emergency Management Agency (FEMA) proposed the whole-community concept and began implementation in 2010. (Source: FEMA/Bill Koplitz.)

can ultimately be leveraged during a disaster. Likewise, a community must foster relationships with community leaders from all sectors—government, nonprofit, and private. The final two characteristics are perhaps the most challenging. Specifically, communities must empower local action and strengthen social infrastructure, networks, and assets. Unfortunately, as discussed in earlier chapters, the current trends indicate that government is centralizing rather than decentralizing power in most areas [7]. Likewise, there is limited (if any) funding for social infrastructure programs to address social needs such as mental health and human wellness.

In addition to the challenges to fund social infrastructure programs, many communities also struggle to support the aging physical infrastructure that supports their communities and is increasingly a risk to local quality of life. For example, the I-35 bridge collapse in Minneapolis, Minnesota, in 2007 killed 13 and injured 145 and pushed this issue into the public consciousness [8] (see Figure 11.2). Unfortunately, public infrastructures such as bridges, sewer systems, and utilities are extremely costly to implement and equally costly to maintain. With shrinking government budgets and a growing number of issues needing public financial support, the financial expenditures on infrastructure have only grown slowly (approximately 2% per year) over the last half-century.

Figure 11.2 Incidents like the 2007 Minneapolis bridge collapse make the public aware of local struggles to maintain aging physical infrastructure in many communities. (Source: FEMA/Todd Swain.)

Unfortunately, the American Society of Civil Engineers (ASCE) released a report stating that this level of funding is insufficient to address the degeneration of these infrastructure systems. Specifically, the ASCE report estimated that it would take $1.6 trillion to upgrade current infrastructure. Similarly, reports from various government agencies estimated that upgrade needs would include $225 billion for roads, $202 billion for wastewater treatment, $72 billion for waterways, $18 billion for airports, $11 billion for drinking water treatment, $10 billion for dams, and $127 billion for schools [9]. Without these upgrades, the risk to these systems increases exponentially; however, clearly the funds to immediately fix the infrastructure are unavailable in most communities. This is a critical element for emergency managers and community leaders to incorporate future sustainability strategies such that the risk is addressed in ways that are culturally, socially, and economically viable.

Given the whole-community approach, sustainability is inherently a politically sensitive topic. For example, nearly 200 mayors and municipal executives from throughout the United States have joined a national campaign called Resilient Communities for America [10]. By signing the agreement, each community commits to creating more resilient cities, towns, counties, etc. built to sustain natural and human-caused risk. Specifically, the campaign identifies extreme weather, energy security, aging infrastructure, and ongoing economic uncertainty [11]. This campaign represents one of the few national movements with actual widespread political support that openly and aggressively focuses on current and future risks from the stated threats and hazards. The campaign provides numerous national and local examples of sustainable risk controls—particularly from climate change impacts (which will be discussed later in this chapter).

However, like many forward-leaning issues—particularly related to disaster management—sustainability does have some political opposition. Because sustainability reflects the current readiness for issues and potential consequences that may (or may not) happen in the future, some political leaders simply do not consider sustainable decisions a priority in current social and financial conditions. There is also concern—particularly among conservative politicians—about the impact of sustainability measures on private development and business success within a given community. Some leaders even approach sustainability concerns as lacking "environmental imperative" and see these measures as anticapitalistic [12]. This view may be inadvertently supported by the fact that a 2014 survey found that nearly 75% of surveyed businesses had not fully implemented sustainable practices [13]. As such, the support of sustainability tends to be partisan with two extreme views of sustainability on the horizon.

In Other Words...Political Objections to Sustainability

Objectors indicate that they are too busy, that the sustainability initiatives are not aligned with the organization's strategic objectives, that the initiatives are too difficult to implement, that there is no pressure from either leadership and/or customers/clients to make these changes, that they've already heard too much about this type of thing, that they don't want to risk their own reputation on unproven ideas, or that ultimately, they'd rather just stick with business as usual.

~Toronto Sustainability Speaker Series [12]

Given the highly partisan nature of these issues, it is difficult to ascertain public perception as it is related to sustainability within communities. Various studies and surveys have been conducted and often show a public desire to build sustainable communities, but simultaneously a very shallow understanding or comprehension of the actual desired results. Much like the discussion earlier in this chapter, sustainability is a very broad term and in some ways is abstract or minimally a true futurist type of issue. Consequently, the general public often defers to those opinions projected by liberal or conservative politicians as well as the perspective of traditional media outlets (which can be equally biased or partisan). This widely distorts the views of sustainability among the general public, which results in the future of sustainability being based more on the strength of public relations than on the need for or importance of building sustainable communities.

ENVIRONMENTAL FACTORS AND CONDITIONS

One of the driving factors for sustainability is the various environmental conditions that surround communities and ultimately impact quality of life. This includes the conditions which create natural hazards and ultimately the broader environmental changes that are potentially exacerbating normal risk within a community. This section of this chapter will discuss these changes with a particular focus on the prevention, control, impact, and consequences of climate change and other related environmental conditions. However, much like the broader issue of sustainability, these environmental factors are not completely understood, applied, or accepted.

Two of the most widely discussed environmental impacts relate to global warming and climate change. The US Environmental Protection Agency (EPA) defines climate change as "any significant change in the measures of climate lasting for an extended period of time" [14]. These measures of climate include temperature, precipitation, wind patterns, or other more targeted measurements (see Figure 11.3). These changes are typically associated with human activities that, since the Industrial Revolution, have released large amounts of carbon dioxide and other greenhouse gases into the atmosphere. The majority of greenhouse gases come from burning fossil fuels in traditional engines and motors; however, other human behaviors such as deforestation, industrial process, and certain agricultural practices can also impact the gas levels and related consequences [14]. While climate change is the broadest and most inclusive environmental impact term, it is often used in association (or even interchangeably) with global warming. Specifically, global warming refers to the "recent and ongoing rise in global average temperature near Earth's surface" due to the greenhouse gas phenomenon already mentioned [14].

Figure 11.3 The US Environmental Protection Agency (EPA) measures climate change through monitoring temperature, precipitation, wind patterns, and other environmental factors. (Source: FEMA/Casey Deshong.)

In Other Words...Impact of Greenhouse Gases

Greenhouse gases act like a blanket around Earth, trapping energy in the atmosphere and causing it to warm. This phenomenon is called the greenhouse effect and is natural and necessary to support life on Earth. However, the buildup of greenhouse gases can change Earth's climate and result in dangerous effects to human health and welfare and to ecosystems.

~US Environmental Protection Agency [14]

The average temperature of Earth has risen by 1.4°F over the past 100 years and is projected to rise another 2°F to 11.5°F (depending on the source and research) over the next 100 years (see Figure 11.4). While these rises seem insignificant on the surface, environmental scientists are alarmed because even minor temperature changes can lead to exponential shifts in the overall climate conditions mentioned earlier. For example,

Figure 11.4 The average temperature of Earth has risen slowly over the past 100 years. (Source: NASA.)

many regions throughout the world have noted changes in rainfall, which has led to more floods or droughts (depending on the shift). Likewise, oceans and glaciers have shown increasing acidity and melting ice caps; this has shown some sea level rising, which has again increased the possibility of severe flooding in coastal communities throughout the world [14].

The concepts of climate change and global warming are not necessarily new to scientists, but have only gained significant traction with government and civil leaders over the last several decades. For example, in 1861 Irish physicist John Tyndall was the first scientist to show the greenhouse gas effect [15] (see Figure 11.5). By 1896, Swedish scientists were postulating that industrial age coal burning would enhance the greenhouse gas effect, but ironically proposed positive and beneficial uses to the world from this discovery. By 1927, with the rise in the development and use of fuel-burning automobiles, carbon emissions from fossil fuel burning reached one billion tons per year. Using documentation from 147 global weather stations, a British engineer named Guy Callendar revealed in 1938 that both temperature and carbon dioxide levels had risen over the previous century. While widely dismissed at the time, this correlation later became known as the Callendar effect. By 1965, US President Lyndon B. Johnson appointed a special advisory committee that ultimately warned

Figure 11.5 In 1861, Irish physicist John Tyndall was the first scientist to show the greenhouse gas effect. (Source: Smithsonian Institute.)

that the greenhouse gas effect was of "real concern" [15]. It was not until 1975 that a US scientist named Wallace Broecker first used the term "global warming" in a scientific paper. By 1988, the United Nations created the Intergovernmental Panel on Climate Change (IPCC), which produced its first climate assessment report by 1990, concluding that temperatures were rising and human activities were most likely the cause. By 2013, the IPCC had released five climate assessments with the final report stating that scientists were 95% certain that humans had been the "dominant cause" of climate change since the 1950s [15].

In addition to the historical context of how the concept of climate change came into existence, it is also important to understand the context of public and government response to these developments. Specifically, in 1997 numerous developed nations pledged to reduce greenhouse gas emissions by an average of 5% by 2008. This so-called Kyoto Protocol identified a wide variety of targeted reductions in individual countries. In the case of the United States, the US Senate immediately declared that it would not ratify the treaty. By 2001, US President George W. Bush officially removed the United States from the Kyoto process and the targeted reductions. By 2005, the Kyoto Protocol became international law for those countries still within the agreement. Three years later incoming US President Barack Obama pledged to "engage vigorously" with the rest of the world on climate control, but did not recommit to the Kyoto standards [15]. The involvement and impact of the United States was and continues to be one of the most contentious issues related to the future of climate change as leaders throughout the world argue that as a major world provider the United States has the moral and practical obligation to reduce climate change.

Much like the Callendar effect discussed earlier, which showed a correlation between temperature rises and carbon dioxide levels, it is also important to note a similar correlation between the occurrences of major disasters and the rise of carbon dioxide over the last half of a century. Specifically, from 1953 to 2011 major disaster declarations in the United States averaged roughly 35 per year, with a slow, but steady increase per year over that time frame. Likewise, the number of declarations issued per decade also increased with a particular spike in the last two decades. In those decades there were 46 and 64 disaster declarations, respectively. This trend is not limited to major disaster declarations as emergency declarations have also increased over that same period of time [16]. While correlation is not definitively causation (as there are numerous other characteristics that impact risk and disaster occurrence), it is highly possible that the greenhouse gas effect and related climate change may be significantly increasing global risk.

These trends are consistent with data available through the International Disaster Database. According to those data, reported natural disasters for the first half of the twentieth century were relatively constant from year to year; however, starting in 1950 the number exponentially grew from year to year without peaking until around 2000 [17]. Likewise, the reported damage from international natural disasters rose astronomically as well, as it moved from only a few billion dollars per year to nearly $100 billion on average by 2011 [18]. However, much like the United States data, this correlation shows strong but not necessarily absolute causation. For example, the number of reported natural disasters in the International Disaster Database actually began to fall after it peaked in 2000 and continued to fall for the next decade [17]. While the overall trends are clear, recent developments (and future trends) may ultimately need additional consideration.

Climate change reports indicate a strong correlation between the human impacts and changes in weather patterns. For example, scientists have observed an increase in the number of heat waves in North America, Europe, Africa, and Asia [2]. For example, there were more than 700 heat-related deaths in Chicago in 1995 and as many as 35,000 in Europe during the summer of 2003 [19,20] (see Figure 11.6). While the US Centers for Disease Control and Prevention (CDC) report that approximately 660 people die each year from heat-related illnesses, the trends are certainly of concern. However, one projection determined that heat-related deaths increased by a factor of 10 when compared to the last century [19]. Given this projection, the number of heat-related deaths in the United States would rise to more than 2,000 per year by 2057 [21].

Additionally, there are numerous projections of the impact of climate control that, while more abstract, may ultimately be just as impactful. Specifically, scientists have projected that climate change and corresponding increases in global temperatures will lead to increasing violence, increases in global food prices, decreases in available groundwater, and reductions in individual and community wealth. While numerous studies have shown a positive correlation between violent crime and increasing temperatures, this correlation is murky as there are numerous other conditions, such as time of day and opportunity that also show connectivity [22]. Likewise, global food prices are projected to rise between 3% and 84% by 2050 due to the warmer temperatures and changes in weather patterns which have and will continue to alter

Figure 11.6 There were more than 700 heat-related deaths in Chicago in 1995. (Source: FEMA/Carolyn Deming.)

growth production [2]. However, it is unclear whether these costs will be evenly distributed or more impactful in areas that are already economically distressed.

In addition to the potential increases in violence, another major concern, particularly in developing countries, is the impact of climate change on the availability of groundwater. Specifically, one study indicated that approximately 33% of the world's population will be impacted by decreases in groundwater supplies by as much as 10% by 2080 [2]. Moreover, the World Health Organization (WHO) has projected that 605 million people will be without improved drinking water and 2.4 billion will lack access to sanitation facilities if this trend continues [23]. Unfortunately, these projections are astronomically far out in the future and often only consider the amount of water entering the groundwater system and whether or not it is subject to human consumption. By taking this limited consideration, the utilization of groundwater for terrestrial vegetation and as part of large water cycles is under-considered [24]. Given this perspective, this issue (like sustainability mentioned earlier) is often difficult to get full acceptance of given more immediate hazards, threats, and community wide risks.

Another commonly discussed environmental impact from climate change is the melting of global ice caps. For example, in early 2002, a section of the Larsen B ice shelf in Antarctica (a section larger than the state of Rhode Island) collapsed due to melting by as much as 40% during the previous decade. Likewise, the Arctic ice thickness has decreased by roughly 9% per decade since the 1960s. Scientists at the US Center for Atmospheric Research have even predicted that the Arctic could be ice-free in the summer by 2040 if the rate of climate change and related global warming continues unchecked. As a result, the IPCC has stated that sea levels have already risen by 4 to 8 inches in the past century and will rise another 10 to 23 inches by 2100 [25]. The US EPA estimates that approximately 26,000 square kilometers of land will be lost by every foot of sea level rise [26]. While these estimates are presented as a wide range over a long period of time, the general consensus among scientists is that these projections are accurate.

These changes create the potential for significant shifts in the risk profile in coastal communities. More than 600 million people in the world live in coastal areas that are less than 30 feet above sea level and two-thirds of the world's cities with populations over five million are located in these high-risk areas. If climate change goes unabated, these high-risk areas will increasingly be impacted by coastal erosion, coastal plain flooding, salinization of soils, and loss of habitat for fish, birds, and other wildlife [26]. These changes in turn create direct hazards such as localized or regional flooding and erode natural barriers that serve as mitigation to these types of risk. Additionally, local community planners and leaders will have to reconsider future development in coastal communities as well as maintenance and mitigation of current properties. Specifically, both 100-year and 500-year floodplains are projected to change in conjunction with the actual sea level rises. In the United States, a change to the 100-year floodplain in conjunction with current estimates from IPCC would result in approximately three to six million more homes falling into these categories with estimated flood damages increasing by 36% to 58% [27].

According to NASA, there are over 200 scientific organizations that publicly support the concept of climate change. Moreover, approximately 97% of scientists—across various disciplines—support the foundations of climate change [28]. Clearly, the science community overwhelmingly supports the science of climate control. However, there is significant partisan and some public objection to climate change. That is not to say there is doubt about or objection to the science as it is unclear what percentage of the public understands the validity of any scientific data or corresponding arguments. However, most objections seem to be oriented around economic equity. Climate change skeptics often argue that if the United States fully engaged in climate change mitigation strategies, it would cause a significant economic strain on an already fragile economic system. It is argued that this economic strain is not equally felt throughout the world as developing countries would not have to limit climate impact as much per capita and thus would not be as economically impacted. As such, in a poll of 39 countries, an average of 54% of international respondents identified global climate change as a major threat to their respective countries while only 40% of Americans responded likewise [29].

In Other Words...Sophisticated Objection to Climate Change

Those who make the sophisticated objection acknowledge that climate change is a serious problem and that the world's nations should be doing something about it. They contend, however, that unilateral action by any country, including the U.S. will impose significant costs without producing significant benefit.... Unilateral action by the U.S. would do nothing about the global stock and little about the global flow [of greenhouse gas emissions]. China is now the biggest greenhouse-gas emitter on earth and in developing nations, emissions are growing at an extraordinary rate. The sophisticated objection is that if the U.S. takes actions on its own, it will impose costs on the American people without seriously addressing the climate problem.

~Cass R. Sunstein [30]

MITIGATION AND ADAPTATION

Given that climate change is widely accepted and most objections are philosophical rather than scientific, it is important for community leaders and emergency managers to identify strategies that reduce the direction and cascading risk from global climate change. Embracing these changes is slow and has not yet been fully embraced. For example, FEMA recently changed its hazard mitigation guidance to include recommendations for "current existing regulatory requirements [and] to consider the risk of 'future events' including consideration of the effects of climate change on disaster risk." However, even with the strong planning language, FEMA did not mandate these changes [31]. While there is some public and professional discussion among emergency managers about the best way to consider the local effects of global climate change, limited (if any) formal planning or preparedness is occurring outside of those actions already addressing natural hazards affected by climate change (e.g., tornadoes).

In Other Words...Planning for Climate Change by Default

Although emergency planners may not be purposefully integrating climate change into their plans, they are already dealing with it by default. What the climate is doing is manifesting itself in weather patterns that are changing...[and] getting more intense, more frequent, and emergency response personnel are at the front lines of that.

~Missy Stults [32]

While limited local, regional, or federal level emergency managers have targeted programming, some global (or perhaps national in some areas) programming or concepts are attempting to consider these issues. The most traditional approach is the concept of adaptation. According to the US EPA, adaptation refers to the efforts by a society or ecosystem to adjust to current and future climate change, which can be manifested through proactive or opportunistic activities. These protective activities include all activities or actions that attempt to protect communities from the impact of climate changes. Similarly, opportunistic activities take advantage of any advantages or beneficial effects created by climate change [33]. These adaptation processes are not new as humans have long adapted to environmental changes—both naturally occurring and those caused by human movement into more hazardous areas. Specific examples of adaptation include the promotion heat-tolerant agricultural crops, implementation of early-warning systems, improving water storage and treatment, and the planting of trees to moderate urban heating patterns [33]. This type of approach is supported and directed by the US Department of Homeland Security in compliance with a 2009 executive order intended to create a national climate change adaptation strategy [34].

While adaptation is widely applied, some outspoken environmentalists like former US vice president Al Gore have stated that adaptation is no longer the most effective strategy for global climate change. Gore argues that there are both financial and philosophical challenges to adaptation. Specifically, he argues that any resources spent on adaptation simply take away from actual mitigation efforts. Moreover, he stresses that the consequences of global climate change are already occurring and if a more aggressive mitigation strategy (rather than simply acceptance and adaptation) is not applied, those consequences will be so devastating that adaptation ultimately will have no effect on reducing the impacts [23]. Likewise, other climate change ethicists have charged that climate change adaptation is too focused on contemporary economics and not enough about "twenty-first century virtues" such as humility, moderation, simplicity, and conservation [35].

Consequently, there is a growing consensus that there needs to be a formal shift away from adaptation strategies toward climate change mitigation. This type of mitigation specifically refers to efforts to cut or prevent the emission of greenhouse gases or the ultimate removal of those gases from the atmosphere. Broadly speaking, climate change mitigation strategies include the creation of new technologies, identification and usage of clean energy sources, changing people's behaviors, and making older technology more energy efficient [36]. For example, there has been a significant public and partisan discussion about energy usage, which has traditionally been heavily based on fossil fuels. Unfortunately, these fossil fuels require significant human, financial, and ultimately environmental resources to move them from raw material to finalized products serving hundreds of different industries.

To minimize the environmental impacts and related climate change, energy-sector climate mitigation has primarily focused on supporting alternative sources of energy provision and creating financial incentives (or disincentives) to use traditional sources that are less efficient and more impactful to the environment. For example, since 2010 more than $211 billion has been invested in US renewable energy sectors with a particular focus in emerging economies [37]. This commitment has shifted the total energy generation of alternative energy sources to 8.5% of the total energy market [36]. Other methods have included the capture and "scrubbing" of carbon dioxide from exhaust systems in traditional power generation systems (e.g., coal). The last and perhaps most politically sensitive approach is the so-called cap-and-trade process that utilizes carbon "credits" to limit overall carbon dioxide emissions by various energy production companies. These carbon credits operate in emission-trading markets that operate differently from country to country based on Kyoto Protocol standards for each country. In theory, if a company's carbon emissions fall below a set allowance, the company is allowed to sell the difference (the credit) to other companies that exceed their limits. The international market for carbon credits has experienced highs and lows (like any economic market), but annual transactions have exceeded more than $60 billion worldwide [38]. However, as of 2013, the total valuation in these carbon markets is overwhelmingly European as only 12% of all transactions are found outside that geographic area [39].

Unfortunately, there are significant objections to some of these mitigation strategies. Specifically, the majority of surveyed Americans object to the concept of carbon credits or any form of taxation of traditional (and environmentally less efficient) forms of energy [40]. Likewise, ethicists and economists question the effectiveness of carbon trading systems. These systems can create loopholes to allow major businesses to evade environmental responsibilities while others have questioned whether carbon trading creates "distributional justice." Lastly, other environmental commentators have simply questioned the actual effectiveness of carbon trading on actually reducing emissions and mitigating the effects of climate change [41]. These systems are clearly not developed enough to truly understood effectiveness at actually mitigating climate change.

Regardless, both adaptation and mitigation are difficult to fully apply at national or international levels, much less in local environments. Consequently, there is a significant question as to how emergency managers should address the future risk of climate change. While various aspects of risk management and disaster response and recovery are politicized, the issues of sustainability and climate control fall significantly along partisan and economic lines. Developed countries are often expected to take a significant lead on risk reduction strategies; however, this is not universally applied or accepted. With the reality that localized hazards and risk have and will continue to impact communities, it is prudent for most emergency managers to continue to monitor and observe current trends related to climate, but primarily to continue the planning, preparedness, and resource management already being leveraged to address the hazards, threats, and risks more clearly understood in a given area.

REFERENCES

1. Almasy, Steve. (2014). "John Kerry: Climate Change as Big a Threat as Terrorism, Poverty, and WMDs." CNN. http://www.cnn.com/2014/02/16/politics/kerry-climate/. Accessed August 16, 2014.
2. "Climate Report: Warming Is a Big Risk to People." *NBC News*. http://www.nbcnews.com/science/environment/climate-report-warming-big-risk-people-n60611. Accessed August 16, 2014.
3. Mileti, Dennis S. and Peek-Gottschlich, Lori. (2000). "Hazards and Sustainable Development in the United States." Colorado State University. http://disaster.colostate.edu/Data/Sites/1/cdra-research/mileti-peek2001.pdf. Accessed August 17, 2014.
4. Mileti, Dennis. (1999). *Disasters by Design: A Reassessment of Natural Hazards in the United States*. New York: A. Joseph Henry Press.
5. Chauvet, Regine Laurence. (2012). "Sustainability and the Emergency Manager: Do They Mesh?" North Dakota State University. http://www.ndsu.edu/fileadmin/emgt/Chauvet_SustainabilityThesis_Defense_GradSchoolEdits.pdf. Accessed August 18, 2014.
6. "Whole Community." (2014). US Federal Emergency Management Agency (FEMA). http://www.fema.gov/whole-community. Accessed August 19, 2014.
7. "A Whole Community Approach to Emergency Management: Principles, Themes, and Pathways for Action." US Federal Emergency Management Agency (FEMA). http://www.fema.gov/media-library-data/20130726-1813-25045-0649/whole_community_dec2011__2_.pdf. Accessed August 19, 2014.
8. Doyle, Pat. (2014). "I-35W Bridge Is Aging, Seven Years after Collapse." *Star Tribune*. http://www.startribune.com/politics/statelocal/268746561.html. Accessed August 20, 2014.
9. "Aging Infrastructure: Issues, Research and Technology." (2010). US Department of Homeland Security. http://www.dhs.gov/xlibrary/assets/st-aging-infrastructure-issues-research-technology.pdf. Accessed August 20, 2014.
10. "Our Communities at Risk." (2014). Resilient Communities for America. http://www.resilientamerica.org/our-communities-at-risk/. Accessed August 19, 2014.
11. "Free Resources for Local Government." (2014). Resilient Communities for America. http://www.resilientamerica.org/how-we-build-resilience/resources-for-local-government/. Accessed August 19, 2014.
12. "Handling Objections: A Professional's Guide to Overcoming Objections to Sustainability Adoption and Implementation." (2012). Toronto Sustainability Speaker Series. http://ecoopportunity.net/wp-content/uploads/2012/04/TSSS-Objection-Discussion-Paper-April-2012.pdf. Accessed August 19, 2014.

13. "Online Poll Finds That 75% of All Companies Still Haven't Embedded Sustainability as Part of Their Employee Culture." (2014). CSRWire. http://www.csrwire.com/press_releases/36755-Online-Poll-Finds-That-75-of-Companies-Still-Haven-t-Embedded-Sustainability-as-Part-of-Their-Employee-Culture. Accessed August 19, 2014.
14. "Climate Change: Basic Information." (2014). US Environmental Protection Agency (EPA). http://www.epa.gov/climatechange/basics/. Accessed August 21, 2014.
15. "A Brief History of Climate Change." (2013). BBC. http://www.bbc.com/news/science-environment-15874560. Accessed August 21, 2014.
16. Lindsay, Bruce R. and McCarthy, Francis X. (2012). "Stafford Act Declarations 1953–2011: Trends and Analyses, and Implications for Congress." Congressional Research Service. http://fas.org/sgp/crs/homesec/R42702.pdf. Accessed August 21, 2014.
17. "Natural Disasters Reported 1900–2011." (2012). EM-DAT. http://www.emdat.be/sites/default/files/Trends/natural/world_1900_2011/eveyr2_view.jpg. Accessed August 21, 2014.
18. "Estimated Damage Caused by Reported Natural Disasters." (2012). EM-DAT. http://www.emdat.be/sites/default/files/Trends/natural/world_1900_2011/damyrTempl_view.jpg. Accessed August 21, 2014.
19. Bienkowski, Brian. (2013). "Heat Waves in Eastern US Will Become Deadlier, Study Says." *Environmental Health News*. http://www.environmentalhealthnews.org/ehs/newscience/2013/11/heat-wave-mortality-in-eastern-us. Accessed August 22, 2014.
20. Bhattacharya, Shaoni. (2003). "European Heatwave Caused 35,000 Deaths." *New Scientist*. http://www.newscientist.com/article/dn4259-european-heatwave-caused-35000-deaths.html#.U_eLxfldWSo. Accessed August 22, 2014.
21. Foster, Joanna M. (2013). "Deaths from Heat Waves May Increase Ten Times by Mid-Century." *Climate Progress*. http://thinkprogress.org/climate/2013/11/08/2917691/heat-waves-ipcc-mortality. Accessed August 22, 2014.
22. Akst, Jef. (2011). "Do Heat Waves Spur Violence?" *The Scientist*. http://www.the-scientist.com/?articles.view/articleNo/30906/title/Do-Heat-Waves-Spur-Violence-/. Accessed August 22, 2014.
23. Gore, Al. (2013). *The Future: Six Drivers of Global Change*. New York: Random House.
24. Clifton, Craig et al. (2010). "Water and Climate Change: Impacts on Groundwater Resources and Adaptation Options." Water Partnership Program. http://www.groundwatergovernance.org/fileadmin/user_upload/groundwatergovernance/docs/Thematic_papers/GWG_Thematic_Paper_12.pdf.
25. "The Consequences of Global Warming on Glaciers and Sea Levels." (n.d). Natural Resources Defense Council. http://www.nrdc.org/globalwarming/fcons/fcons4.asp. Accessed August 23, 2014.

26. "Oceans and Sea Level Rise." (2010). Climate Institute. http://www.climate.org/topics/sea-level/. Accessed August 23, 2014.
27. "Projected Impact of Relative Sea Level Rise on National Flood Insurance Program." (n.d.). Rising Sea. http://papers.risingsea.net/Flood-Insurance.html. Accessed August 23, 2014.
28. "Consensus: 97% of Climate Scientists Agree." (n.d). National Aeronautics and Space Administration. http://climate.nasa.gov/scientific-consensus/. Accessed August 23, 2014.
29. "Climate Change: Key Data Points from Pew Research." (2014). Pew Research Center. http://www.pewresearch.org/key-data-points/climate-change-key-data-points-from-pew-research. Accessed August 25, 2014.
30. Sunstein, Cass R. (2013). "U.S. Should Act Unilaterally on Climate Change." *Bloomberg View.* http://www.bloombergview.com/articles/2013-01-23/u-s-should-act-unilaterally-on-climate-change. Accessed August 24, 2014.
31. Hammer, Becky. (2014). "FEMA Promises to Revise State Planning Guidance to Require Consideration of Climate Change." NRDC Switchboard. http://switchboard.nrdc.org/blogs/rhammer/fema_promises_to_revise_state.html. Accessed August 26, 2014.
32. Pittman, Elaine. (2010). "Emergency Managers Warm to the Idea of Climate Change." *Emergency Management Magazine.* http://www.emergencymgmt.com/disaster/Emergency-Managers-Climate-Change.html. Accessed August 26, 2014.
33. "Adaptation Overview." (2013). US Environmental Protection Agency (EPA). http://www.epa.gov/climatechange/impacts-adaptation/adapt-overview.html. Accessed August 26, 2014.
34. "Department of Homeland Security Climate Change Adaptation Roadmap." (2012). US Department of Homeland Security. https://www.llis.dhs.gov/content/department-homeland-security-climate-change-adaptation-roadmap. Accessed August 26, 2014.
35. Gardiner, Stephen. (2004). "Ethics and Global Climate Change." *Ethics Journal.* http://www.jstor.org/stable/10.1086/382247. Accessed August 27, 2014.
36. "What Is Climate Change Mitigation?" (2014). *BBC News.* http://www.bbc.com/news/science-environment-26980837. Accessed August 27, 2014.
37. "Climate Change Mitigation—Energy." (2014). United Nations Environment Program. http://www.unep.org/climatechange/mitigation/Energy/tabid/104339/Default.aspx. Accessed August 27, 2014.
38. Stillman, Jessica. (2008). "What Is Carbon Credit?" *CBS News.* http://www.cbsnews.com/news/what-is-carbon-credit/. Accessed August 27, 2014.
39. Upton, John. (2014). "Carbon Trading Is Booming in North America, No Thanks to U.S. and Canadian Governments." *Grist.* http://grist.org/news/carbon-trading-is-booming-in-north-america-no-thanks-to-u-s-or-canadian-governments/. Accessed August 27, 2014.

40. Koch, Wendy. (2014). "Poll: 60% Back Carbon Tax If Used for Renewables." *USA Today*. http://www.usatoday.com/story/money/business/2014/07/21/poll-on-carbon-tax-finds-mixed-support/12950925/. Accessed August 28, 2014.
41. Hepburn, Cameron. (2011). "Carbon Trading: Unethical, Unjust and Ineffective?" Cameron Hepburn Blog. http://www.cameronhepburn.com/research/publications/academic-papers/carbon-trading-unethical-unjust-and-ineffective/. Accessed August 28, 2014.

12

Diversity, Globalization, and Extremism

> Extremism stifles true progression in all fields of human advancement... their language binds people together, but only by stirring the darkest excesses of the soul.
>
> ~Daniel S. Fletcher [1]

DIVERSITY AND SOCIOECONOMIC FACTORS

Diversity is one of the many "hot-button" topics discussed in this book. Diversity relates to the interconnectedness of social, cultural, political, and ultimately economic factors within communities of various sizes. Likewise, with the surge of technologies and the advent of mobility and interconnected systems, the world has become a much smaller place with social and economic interactions becoming increasingly diversified. Unfortunately, the intentional and unintentional blending of these components has not always occurred easily. Various forms of dissension—including extreme views of politics, religion, and culture—are occurring with increasing frequency and they create increasing community risk with often unpredictable (or not previously occurring) conditions and results.

The importance of incorporating diversity in emergency management and homeland security planning and preparedness has significant historical support. For example, there was a disproportionate level of impact and costs to racial and ethnically diverse communities who were impacted by Hurricane Katrina. Likewise, other national events, like the H1N1 influenza pandemic of 2009, showed significantly higher rates of illness, hospitalization, and death in more diverse populations when compared with Caucasian populations [2] (see Figure 12.1). While these statistics become clearer after disasters, they are not inherently related to disasters. They reflect cultural and social divisions that exist before, during, and after disruptive events and are routinely discussed among partisan lines with significant questions about what role government should play in addressing these diversity issues.

Unfortunately, surveys of emergency management and preparedness websites indicate there is poor acceptance and application of diversity within emergency planning. Specifically, less than 40% of those surveyed acknowledged the importance of preparedness as it concerns diversity among the served population. Likewise, only 13% of surveyed websites provided information, materials, or publications that focused predominantly on preparing diverse communities for community risks and methods of preparedness. Lastly, only half of the websites made any mention of racial or ethnic diversity within the served area [3]. While racial and ethnic diversity varies from community to community, a wide variety of opinions, attitudes, races, ethnicities, faith systems, economic stabilities,

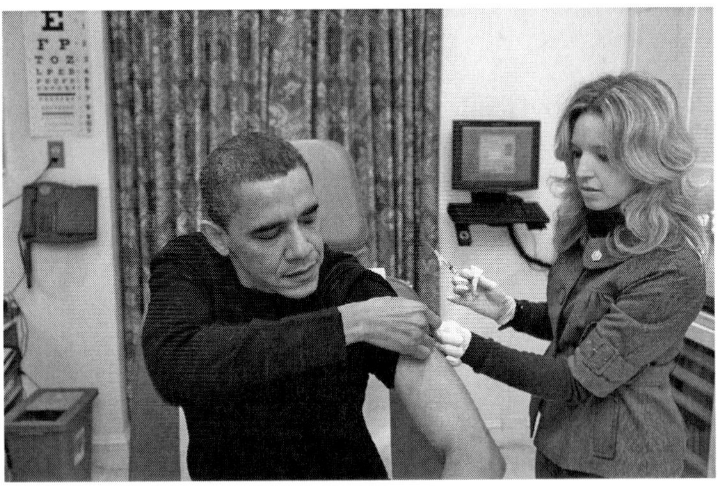

Figure 12.1 President Obama was vaccinated for the H1N1 influenza virus in 2009. (Source: White House/Pete Souza.)

political ideologies, and other personal factors do create diversity. Understanding this dynamic is one of the biggest hurdles facing the emergency management and homeland security community. Because these elements of diversity are often poorly understood, identified, and prepared for, common barriers are created which impact how effectively a community can prepare for disruptive incidents and disasters.

In Other Words...Definition of Diversity

The concept of diversity encompasses acceptance and respect. It means understanding that each individual is unique and recognizing our individual differences. These can be along the dimensions of race, ethnicity, gender, sexual orientation, socio-economic status, age, physical abilities, religious beliefs, political beliefs, or ideologies. It is the exploration of these differences in a safe, positive, and nurturing environment. It is about understanding each other and moving beyond simple tolerance to embracing and celebrating the rich dimensions of diversity contained within each individual.

~The City University of New York (CUNY) [4]

The first of these common barriers is social and economic factors. According to Northwestern University, socioeconomic factors measure an individual's or family's economic and social position based on education, income, and occupation. Consequently, measures of income and relative poverty (if applicable) as well as income and employment and educational conditions, such as level of achievement and occupation, are critical to understanding how these factors impact disaster response and recovery [5]. Research literature has routinely established that preexisting socioeconomic characteristics significantly impact the ability of individuals and communities to cope with the impacts of disasters and other disruptive events. For example, one researcher stated that "people's needs are grounded in the nature of their lives before the disaster began; specifically in their employment status, financial resources, social supports, legal entitlements and housing situation" [5]. For example, the 1994 Northridge (California) earthquake revealed that local Latino communities that were predominantly limited to low-wage agricultural employment before the earthquake were highly impacted

Figure 12.2 During the 1994 Northridge (California) earthquake, the Latino community was disproportionately impacted due to pre-existing issues. (Source: FEMA/Andrea Booher.)

during the recovery process and were confined to substandard and unsafe housing, which contributed to their inability to recovery quickly and effectively [5] (see Figure 12.2).

The second factor is related to culture and language. In many communities limited language proficiency in the primary localized language (e.g., English in the United States) is a significant challenge to disaster preparedness because these community sectors are less likely to understand emergency messages, warnings, and protective action statements [5]. Unfortunately, most communities lack the resources to adequately provide proactive and reactive translation resources. Moreover, there may be significant cultural components, which are often overlooked or misunderstood by emergency management communities. For example, during the 2011 tornado outbreak in Alabama, the Joint Field Office (JFO) run by the US Federal Emergency Management Agency (FEMA) had difficulty creating postdisaster recovery information for certain communities in rural parts of Alabama that had well-developed (but unusual to other parts of the state) cultural standards as well as long-standing distrust of state and federal government (see Figure 12.3). Consequently, significantly more time and resources were necessary to insert the message effectively into this community and others with strong and pre-existing cultural characteristics.

Figure 12.3 In 2011, in response to a tornado outbreak in Alabama, FEMA had difficulty creating postdisaster recovery information for certain communities in rural parts of Alabama with well-developed cultural standards. (Source: FEMA/ David Fine.)

The third characteristic that highly impacts community emergency preparedness is the level of trust and perceived fairness of government response in more diverse communities. This lack of trust in government actions pervades all aspects of emergency response and recovery, including emergency communications, warning messages, and resource management activities within the communities. Consequently, it reduces the likelihood that appropriate response and protective actions are initiated in a timely manner. This distrust is particularly present in undocumented immigrant communities due to fears of deportation, which compound language proficiency issues already present in many of these communities [5]. Likewise, numerous research studies have shown this distrust in government to be prevalent in racially and ethnically diverse populations as well due to past experiences of real or perceived discrimination [5].

Unfortunately, these impacts are significant as they are becoming increasingly common in communities throughout the world. For example, in the United States the number of undocumented immigrants has risen from 3.5 million in 1990 to nearly 12 million by 2012 [6]. Currently, as much as 60% of these illegal immigrants come from six large and highly industrialized states (New York, New Jersey, Texas, California,

Florida, and Illinois) [7] Approximately six million to seven million of these undocumented immigrants are native to Mexico [8]. This is one of the more challenging components as now language and lack of trust in government become critical to limiting how people will respond during emergencies.

Additionally, some politicians and homeland security advisors have suggested that the immigration influx into the United States has created or may create an increased risk from terrorism. These concerns have recently been inflamed by increased violence near Mexican border towns such as Nuevo Laredo, Reynosa, Juarez, and Tijuana, where it has been speculated that guns, violence, and potential terrorists can move freely back and forth across the shared border. While typically leveraged for political purposes, these concerns have also been inflamed by radical terrorists who have suggested the possibility of smuggling of jihadists across this border. However, there is no public information to suggest a real link between international terrorism and immigration issues in the United States. Interestingly, there have been more documented cases of terrorist plots in the United States involving the US–Canadian border or air travel than traditional movement in and out of the US–Mexican border [9].

These issues are not limited to undocumented immigrant populations as Latinos represent the largest minority group and fastest growing group population sector in the United States. According to the US Census Bureau, Latino (or Hispanic) populations represent approximately 17% of the entire US population and more than half of the population growth over the past decade [10]. Of these identified Latino populations, 78% speak Spanish at home and the majority is heavily dependent on community-based organizations for information and services. Unfortunately, more than 25% of these community members lack a regular health care provider [11]. These characteristics and others like them create cultural norms among these populations that include strong familial ties, affiliation over confrontation, and traditional gender roles [11]. Unfortunately, these communities, like the undocumented immigrants discussed earlier, are often misunderstood by the emergency management community. Formal planning approaches tend to lump all Latino communities and characteristics into one approach, which may inadvertently undermine the effectiveness of creating additional language, cultural, or trust issues [12].

Unfortunately, emergency management planning challenges for Latino and immigrant communities are not the only diverse community that is often poorly prepared for. Specifically, the US Centers for Disease Control and Prevention (CDC) report that African Americans represent approximately 14% of the population of the United States, with this percentage projected to rise above 18.4% by 2060. While African Americans live throughout the United States, approximately 55% live in the southern United States (e.g., Georgia, Alabama, Mississippi, etc.) [11]. Much like the cultural characteristics of the Latino community mentioned earlier, the African American community is also highly developed around certain cultural markers. For example, communication is highly contextualized with the use of indirect cues and shared experiences. Likewise, there is a strong respect for elders and collaborative approaches to problem solving. However, there is long-standing concern over racism and unfounded interference, bias, and harassment from governmental agents (e.g., law enforcement). Moreover, an identifiable portion of the African American community is more likely than other racial or cultural groups to be living in poverty and within a lower socioeconomic threshold [13]. Consequently, traditional approaches to emergency preparedness are not always effective.

In Other Words...Vulnerability of African American Communities

African American communities are disproportionately vulnerable to and impacted by disasters. [This] socio-economic vulnerability is based on multiple factors, including pervasive lack of wealth...[and] oft-compromised quality of housing stock in many...communities.... Yet studies show that African American households are significantly less likely to be prepared for disasters than White Americans. Plus, African Americans are under-represented in disaster response design and implementation. Thus, the systems and protocols established to address disasters often don't fully take cultures and circumstances of African Americans into account, resulting in response mechanisms that fall far short of meeting...needs.

~National Association for the Advancement of Colored People (NAACP) [14]

As alluded to earlier, lower socioeconomic status, poverty, or near-poverty levels can play a significant role in emergency preparedness and acceptance of lifesaving information. While some measures are indicating stabilization in the rate of poverty in the United States at approximately 15%, previous trends have indicated a growing number of individuals in this category. Likewise, there has been a decline in both household income and households with health insurance [15]. This decline in income has created a greater divide between higher and lower socioeconomic levels, with a particular decrease in the middle class. This so-called "Lost Decade" of the middle class has led to 85% of self-described middle class adults stating both a decline in median net worth and an ability to maintain standards of living [16]. While standards of living and definitions of poverty are impacted by macroeconomic trends (e.g., global recession) and political or governmental definitions, it creates significant hurdles for preparedness as both lower and middle socioeconomic classes are less able to maintain desired living conditions, much less the extra processes needed to prepare for disasters or mitigate personal or community risks.

Global poverty is no less challenging. For example, over three billion people worldwide live on less than $2.50 per day and at least 80% of the world's population lives on less than $10 per day (see Figure 12.4). Moreover, according to the United Nations Development Program, more than 80% of the global population lives in countries where income differentials are widening [17]. This differential is analyzed by the Gini coefficient, which measures inequality of income on a scale of 0 to 100, with 0 representing an ideal where all people would make the same income.

Figure 12.4 Global poverty continues to be a challenge, with much of the world living on less than $10 per day. (Source: World Health Organization/Pierre Claquin, MD.)

For example, the United States is rated 35–45, China is rated 30–40, Russia is 20–40, and the United Kingdom is 30–36 [18]. Interestingly, according to Gini measures by the World Bank, no country is currently lower than 25 (Ukraine) or higher than 63 (South Africa) [19]. Much like the United States, understanding the economic limitations will help emergency management professionals plan for future risks and challenges.

In addition to the traditional racial, cultural, and socioeconomic components, part of the consideration of diversity is the acknowledgment that there are different opinions, interests, and health conditions. This broader definition encompasses some emerging trends for emergency managers, including the need to consider mental health trends within local communities. Understanding and observing mental health in communities is critical because "unlike physical injuries, adverse mental health outcomes of disasters may not be apparent, and therefore a systematic approach to…appropriate interventions is required" [20]. For example, according to research published in the *Journal of the American Medical Association* (*JAMA*), up to 40% of individuals involved in emergencies, disasters, and other disruptive events (e.g., active shooters) had pre-existing mental health disorders. Moreover, between 11% and 38% of distressed individuals utilizing shelters or family assistance centers during emergencies or disasters have "stress-related or adjustment disorders" [20].

Pre-existing mental health awareness is not the only element of understanding this type of community component. Disasters and disruptive community events may not only exacerbate pre-existing issues but also trigger various forms of disaster mental health issues. While there are some nongovernmental organizations and specialized government agencies focused on the impact of disasters on mental health wellness, the formalized approach is not widespread. This is particularly noteworthy, as nearly 90% of all people will experience some form of disaster-related trauma in their lifetime [20]. If these postevent mental health issues go untreated, individuals can experience long-term anxiety or depression, inability to care for themselves, suicidal or homicidal thoughts, and abuse of alcohol or drugs as well as domestic violence, child abuse, and elder abuse [21] (see Figure 12.5). Consequently, if emergency management professionals do not better incorporate these considerations into planning for all stages of disaster management, individual problems may cascade into additional community impacts or secondary issues.

The last component of community diversity that is of growing importance is the rise of changes in sexuality and gender roles. As discussed in Chapter 5, the lesbian, gay, bisexual, and transsexual (LGBT)

Figure 12.5 Stress from disasters may exacerbate pre-existing health issues such as the abuse of alcohol. (Source: Paolo Neo.)

communities have increasingly become more active and broadly accepted within communities. However, not all emergency management planning and preparedness have adjusted to follow suit. For example, emergency sheltering operations are often established around the separation of men, women, and families. However, as the definition and roles of gender have shifted and the concept of family has changed, shelter operations are struggling to maintain traditional approaches to protect all those in need of shelter in a fair, equitable, and respectful way. While still a relatively small percentage of most communities, LGBT communities have a clear and collected voice and executive, legislative, and judicial branches of government at all levels are increasingly preserving LGBT rights, which must be better understood by emergency managers in the future to ensure that ethical and legal standards can be fully met.

GLOBALIZATION

In addition to diversity, the current and future trends in globalization will significantly impact emergency preparedness. Unfortunately, globalization is poorly defined and somewhat clichéd when viewed strictly from political and cultural perspectives. However, there are some ways to view and understand globalization. For example, globalization partially reflects the perception that there is a broadening, deepening, and speeding up of international connectedness throughout all aspects of life. This connectedness is all encompassing to include cultural factors, personal and community security, financial markets, and environmental factors, just to name a few. The most important element is how these interconnected components alter quality of life issues such as safety and resiliency [22].

Globalization is characterized by four types of change. First, there is a stretching of social, political, and economic activities across all geopolitical boundaries. Secondly, there is an intensification of the flow of global economic trade, investment, finance, and cultural exchanges. The third type of change reflects growing intensity of interconnectedness through the evolution of global transportation and communication systems. The fourth and final type of change relates to an increase in the intensity of global interactions such that local changes can have global impacts [22]. For perspective, an agricultural example will be considered. The importation of fresh fruit and vegetables increased exponentially in the United States over the last 20 years (see Figure 12.6). While the reasons are complicated, globalization is at the core as "the strong growth in the volume and variety of fresh product imports has allowed U.S. consumers to eat more fruit and vegetables…year-round" [23]. While climate conditions and the perishable nature of fruits and vegetables have traditionally limited the distribution of a wide variety of fresh fruits year-round, the specific and targeted globalization (for economic and quality-of-life issues in this case) has vastly expanded its reach. Conversely, when there is global pressure on sensitive and perishable systems (like fresh fruit), the price and availability throughout the world can be highly impacted with price swings that have long-term global impacts.

This idea is further expanded upon in Thomas Friedman's well-known analysis of globalization, *The World Is Flat*. In the book, Friedman describes 10 so-called "flatteners" that level

> …the global playing field…[and make] it…now possible for more people than ever to collaborate and compete in real time with more other people on more different kinds of work from more different corners of the plan and on a more equal footing than at any previous time in the history of the world [24].

This list of flatteners includes the rise of computer manufacturing, Internet, social media, virtual communities, outsourcing, offshoring, and

Figure 12.6 The cost of fruits and vegetables has increased exponentially in the United States over the last 20 years due to the influences of globalization. (Source: CDC/Eric Grafman.)

so-called "steroid" approaches that combine these various effects [24]. While emergency managers are not typically (nor will they be in the future) responsible for trying to control or influence these types of issues, globalization does have significant impacts to disaster management effectiveness and success.

In Other Words...Negative Impacts of Globalization

The main single cause of increases in poverty and inequality in the 1980s and 1990s was the *"retreat of state."* As states' welfare provision declined, the "ability of working class forces...to resist the negative impacts of globalization on working conditions and employment is severely diminished."

~United Nations [25]

According to a study from the United Nations, globalization causes significant alterations to natural socioeconomic conditions throughout the globe. Consequently, the increasing poverty and economic inequality discussed earlier can be exacerbated, which in turn creates additional vulnerability from natural hazards, particularly in densely populated urban areas [25]. Additionally, some research studies have suggested globalization also reduces the ability of national governments to manage risk. Specifically, deregulation of industrial systems (to support global exchange) pushes control of economic conditions and other factors into the hands of corporations. Because of the economic incentives of business development, the argument has been made that the protection and preservation of environmental conditions is not paramount and therefore creates additional risk from development, operations, and other components [25].

When globalization contributes to increased vulnerability in lower socioeconomic areas, it can be particularly present in the availability and effectiveness of housing and health care systems. For example, according to a study from the Geneva Centre for Security Policy, a recent earthquake in Pakistan damaged or destroyed more than 70% of health facilities and more than 200,000 housing units and damaged nearly another 200,000 units. Likewise, a recent Indonesian tsunami resulted in more than 90% of all hospital staff either missing or dead [26]. Moreover, these types of exacerbated effects are not limited to developing countries. There was significant destruction or damage to the housing populations after Hurricanes Rita and Katrina in 2005 (see Figure 12.7). Specifically, more than 300,000 houses were destroyed and an additional 1.85 million were damaged [26]. Moreover, the vast displacement of individuals significantly hindered the long-term recovery as these individuals were poorly tracked and in some cases never returned to the impacted areas.

These impacts of globalization result in significant policy dilemmas for emergency management communities and related political officials. For example, given the interconnectedness of culture, society, and economies, some individuals have called for a shared responsibility to emergency response and management. This so-called transnational emergency response structure would include functions such as improved warning systems, global communication structures, and other integrated planning approaches [26]. While this type of global emergency response is commonly proposed by the United Nations, it is often poorly executed in a truly equitable and global way. The equitable responsibility of globalization now and into the future will be a challenge for developed and developing countries alike. Much like the environmental impacts

Figure 12.7 There was significant destruction or damage to the housing populations after Hurricanes Rita and Katrina. (Source: FEMA/Hans Pennink.)

discussed in previous chapters, perspective is a critical issue in what is fair, equitable, approachable, and effective toward reducing risk from globalization.

RELIGIOUS EXTREMISM

In addition to the spread of diversity and globalization, a third leg of society, culture, and connectivity must be considered. Specifically, various forms of extremism have gained stronger roots (perhaps inspiration) and wider platforms over the last several years, which have greatly increased their ability to spread messages of hate and division. Previously, divergent or opposite perspectives, values, biases, or other factors have been thrust together or exposed to one another in quick and often drastic ways. For example, corporations and businesses from developed countries often promote and encourage equality among races and gender; however, when those same companies expand operations into less developed areas to utilize raw materials or leverage new markets, those views can contrast significantly to pre-existing caste or cultural systems. While the specific elements and characteristics of extremism will be further considered from an emergency management and homeland security perspective, the rest of this chapter will primarily focus on the political and religious extremist perspectives and groups that have begun to arise both domestically and internationally.

In Other Words....Challenge of Extremist Traits

What is objectionable, what is dangerous about extremists is not that they are extreme, but that they are intolerant. The evil is not what they say about their cause, but what they say about their opponents.

~John F. Kennedy [27]

While there is no universally agreed-upon definition of extremism, it generally reflects taking political or religious ideology to the limits (or extremes) of its application with no regard for repercussions, ramifications, impracticalities, or objections. Because of the extreme nature of these views, they inherently stand in opposition to any opposing view which may be moderate or the opposite extreme. These oppositionist views become the focus of the extremist view which desires to eliminate all opposing views. This attempt at elimination often comes without regard for life, liberty, or human rights [28]. Consequently, some observers have reflected that extremism is "essentially more an issue of style than content" [28]. Extremists' behavior is characterized by a variety of behaviors including character assassination, name calling, sweeping generalizations, advocacy of double standards, use of buzzwords, doomsday thinking, moral superiority, and inclination toward "group think," among others [27]. Likewise, extremists also tend to view their opposition as essentially evil and call for their censorship. Extremists also tend toward emotional responses, intimidation, hypersensitivity, and the use of supernatural rationale, which leads to significant difficulties tolerating ambiguity and uncertainty [27].

In Other Words...Extremism on the "Tails"

Extremists hold political preferences that, in any distribution of opinion, lie in one of the "tails." In other words, their political beliefs are not widely shared even within their own societies...[therefore] extremists currently lack the means or power to obtain their goals.

~David A. Lake [29]

Religious extremism is one form of this behavior that has had significant public safety and emergency management implications. Therefore, it is important to understand how this extremism has developed over the last 30 years. In the early 1980s the US State Department's list of international terrorist groups listed almost zero religious extremist groups. However, by 1998, approximately 15 religious extremist groups were identified by the US State Department. By 1994 and 1995 the number of religious extremist groups had increased to 16 and 26, respectively [30]. This rise was most clearly exemplified in the terrorist attacks in the late 1990s and early 2000s that were perpetrated by the emergent extremist Islamic group called al-Qaeda. These incidents included the 1993 bombing of the World Trade Center, 2001 World Trade Center airplane attacks, 2005 London subway bombings, 2009 "underwear" bomber, and various other acts of religiously influenced terrorism.

While Islamic extremism has been localized or regional issues since the late 1970s, the spread to a global terrorist risk is decidedly more recent. Specifically, as the Cold War ended and many central Asian and Middle Eastern countries destabilized, the advanced weaponry and military knowledge previously controlled and contained by governmental structures began to proliferate across the region, particularly in areas with weak centralized government structures. This was particularly evident in Afghanistan, where smuggling and drug trafficking were exploited by religious extremists to fund a collective focus on terrorism [31]. It is against this backdrop that al-Qaeda was established by Islamic terrorist Osama bin Laden as a transnational network of terrorism to "re-establish the Muslim state" through the overthrow of "corrupt" Islamic government regimes and the removal of American and Israeli influence [31]. Ultimately, this loose affiliation of Islamic extremists was responsible for the bombing of the US Embassy in Nairobi and the *USS Cole*, and that facilitated the multiple attacks on September 11, 2001 (see Figure 12.8).

While the United States and various allies instituted militaristic conflicts in Afghanistan and eventually Iraq to try to eliminate al-Qaeda cells and various supportive loosely centralized governments, the ultimate results yet again created a vacuum of centralized control within society and culture. Much like the post-Cold War effects, Islamic extremism has filled the gap and will continue to be a threat into the future. Specifically, the Islamic State of Iraq and Syria (ISIS) has risen as a semicollected group of extremists attempting to assert control throughout the Middle East [32]. While al-Qaeda was extremely impactful, ISIS is at a whole different level of impact as it openly controls a broader

Figure 12.8 The al-Qaeda network of Islamic extremists was responsible for the bombing of the *USS Cole*. (US Navy.)

geographic range and has better economic stability as an organization (e.g., oil revenue vs. illicit drug trade). Moreover, its religious approach is even more extreme, with online beheadings of Western journalists and public calls for religious conversion or death. One commentary referred to ISIS as "al-Qaeda on steroids" [33].

In Other Words...The Emergence of ISIS

The emergence of terrorism in a new form has sent tremors of anxiety around the globe. ISIS, the self-proclaimed Islamic state and caliphate, has redefined the meaning of terror by combining relentless barbarism with military skill, religious bigotry, unprecedented affluence and the shrewd use of video and social media. In Syria and Iraq, it has captured territory occupied by about 8 million people. If it holds those lands, its rise will be the most significant revision of Middle East geography since national borders were set after the First World War.

~Robert Fulford [33]

Interestingly, within emergency management communities the concept of considering religious extremism and its connection to terrorism is often thought of solely as Islamic threats to developed or non-Muslim countries. However, this assumption is flawed. There is an increasing concern throughout the Muslim world over the rise of groups like ISIS. For example, according to the Pew Research Center, between 75% and 92% of those surveyed in Egypt, Tunisia, and Lebanon were very or somewhat concerned about the rise of ISIS [34]. Moreover, in the United States, the assumption that all acts of terrorism are related to Islamic extremism is also overstated. For example, according to a 2013 study by the Triangle Center on Terrorism and Homeland Security, only 33% of the American deaths from political or religious violence since the September 11 attacks

Figure 12.9 The bombing of the Oklahoma City federal building in 1995 was perpetrated by a white supremacist. (US Department of Defense.)

have come from individuals who identified themselves as Muslim. Moreover, in 2012, all but one of the nine Muslim-related terrorism plots were uncovered in early stages [35]. The threat of Islamic extremist and, by extension, terrorist actions is not the only form of religious extremism that can increase risk in a community.

It is important to note that Islamic extremism is not the only form of religious extremism that can create risk in a given community. There are significant historical examples of violence and terrorism that have been perpetrated by non-Muslim individuals. For example, the shooting massacre at a Wisconsin Sikh temple in August 2012 was fueled by a white supremacist that mistakenly associated the Sikh attire with Muslims. Similarly, the bombing of the Oklahoma City federal building in 1995 was perpetrated by another white supremacist [36] (see Figure 12.9). In the case of the bombing of the federal building, this event represented the most significant act of terrorism in the United States until the September 11 attacks. There are also emergent Christian terror groups, like the Anti-Balaka in the Central African Republic, which are targeting and attacking Muslim refugee camps as part of widespread political tensions in the area [37]. Much like the Islamic extremists discussed earlier, mainline Christian groups have condemned these actions, but in the end the extremist risk exists outside those views or any needed support of mainline or temperate views.

REFERENCES

1. Fletcher, Daniel S. (2014). *Jackboot Britain: The Alternate History—Hitler's Victory and Nazi UK!* Create Space Independent Publishing.
2. "Guidance for Integrating Culturally Diverse Communities into Planning for and Responding to Emergencies: A Toolkit." US Human & Human Services (HHS). http://www.hhs.gov/ocr/civilrights/resources/specialtopics/emergencypre/omh_diversitytoolkit.pdf. Accessed August 29, 2014.
3. Andrulis, Dennis P., Siddiqui, Nadia J., and Gantner, Jenna L. (2014). "Preparing Racially and Ethnically Diverse Communities for Public Health Emergencies." *Health Affairs.* http://content.healthaffairs.org/content/26/5/1269.full. Accessed August 29, 2014.
4. "Definition of Diversity." (n.d.). Queensborough Community College of CUNY. http://www.qcc.cuny.edu/diversity/definition.html. Accessed August 29, 2014.
5. Andrulis, Dennis, Siddiqui, Nadia, and Purtle, Jonathan. (2009). "California's Emergency Preparedness Efforts for Culturally Diverse Communities." Center for Health Equity at Drexel University. http://www.diversitypreparedness.

org/SiteData/docs/Drexel%20California%20Preparedness%20Diversity%20Report/d6c2049cf02d63c8420c95693f50fa92/Drexel%20California%20Preparedness%20Diversity%20Report.pdf. Accessed August 30, 2014.
6. "Following a Recession-Related Decline, US Unauthorized Immigration May Be on the Rise." (2013). PewResearch Hispanic Trends Project. http://www.pewhispanic.org/2013/09/23/unauthorized-immigration/1-3/. Accessed September 1, 2014.
7. "These Six States Were Home to 60% of the Unauthorized Immigrants in the United States." (2013). PewResearch Hispanic Trends Project. http://www.pewhispanic.org/2013/09/23/unauthorized-immigration/2-3. Accessed September 1, 2014.
8. "Most Unauthorized Immigrants Are from Mexico, but Their Number Is Down from Its Peak." (2013). PewResearch Hispanic Trends Project. http://www.pewhispanic.org/2013/09/23/unauthorized-immigration/3-3/. Accessed September 1, 2014.
9. Steward, Scott. (2014). "Examining the Terrorist Threat from America's Southern Border." *Stratfor Global Intelligence*. http://www.stratfor.com/weekly/examining-terrorist-threat-americas-southern-border#axzz3C7Qt7Dv3. Accessed September 1, 2014.
10. Brown, Anna and Lopez, Mark Hugo. (2013). "Mapping the Latino Population by State, County, and City." Pew Research Hispanic Population Trends. http://www.pewhispanic.org/2013/08/29/mapping-the-latino-population-by-state-county-and-city/. Accessed September 2, 2014.
11. "Minority Health." (2014). US Centers for Disease Control & Prevention (CDC). http://www.cdc.gov/minorityhealth/populations/remp/black.html. Accessed September 2, 2014.
12. "Latinos during Emergencies: Cultural Considerations Impacting Disaster Preparedness." (n.d.). University at Albany Center for Public Health Preparedness. http://www.albany.edu/sph/images/Latinos.pdf. Accessed September 2, 2014.
13. Donovan, Susan. (n.d.). "Family Characteristics within a Cultural Framework." Florida Health Department. http://www.floridahealth.gov/alternatesites/cms-kids/providers/early_steps/training/documents/family_characteristics.pdf. Accessed September 2, 2014.
14. "Emergency Management." (2014). National Association for the Advancement of Colored People (NAACP). http://www.naacp.org/pages/emergency-management. Accessed September 2, 2014.
15. Cohn, D'Vera. (2012). "Americans' Views about Poverty and Economic Well-Being." Pew Research Social & Demographic Trends. http://www.pewsocialtrends.org/2012/09/12/americans-views-about-poverty/. Accessed September 3, 2014.
16. "The Lost Decade of the Middle Class." (2012). Pew Research Social & Demographic Trends. http://www.pewsocialtrends.org/2012/08/22/the-lost-decade-of-the-middle-class/. Accessed September 3, 2014.

17. Shah, Anup. (2013). "Poverty Facts and Stats." *Global Issues.* http://www.globalissues.org/article/26/poverty-facts-and-stats. Accessed September 3, 2014.
18. Gore, Al. (2013). *The Future: Six Drivers of Global Change.* New York: Random House.
19. "GINI Index." The World Bank. http://data.worldbank.org/indicator/SI.POV.GINI. Accessed September 3, 2014.
20. North, Carol S. and Pfefferbaum, Betty. (2013). "Mental Health Response to Community Disasters." *Journal of the American Medical Association (JAMA).* http://jama.jamanetwork.com/article.aspx?articleid=1724280. Accessed September 4, 2014.
21. "Disaster Mental Health for Responders: Key Principles, Issues and Questions." (2012). US Centers for Disease Control & Prevention (CDC). http://www.bt.cdc.gov/mentalhealth/responders.asp. Accessed September 4, 2014.
22. Held, David et al. (1999). "What Is Globalization?" *Global Transformations.* http://www.polity.co.uk/global/whatisglobalization.asp. Accessed September 5, 2014.
23. DiMartino, Christina. (2012). "US Fruit and Vegetable Imports Growing Larger by the Year." *The Produce News.* http://www.producenews.com/markets-and-trends/9248-u-s-fruit-and-vegetable-imports-growing-larger-by-the-year. Accessed September 5, 2014.
24. Altermatt, Bill. (2006). "Summary and Excerpts from Thomas Friedman's *The World Is Flat.*" Notes on *The World Is Flat.* http://kursinfo.himolde.no/in-kurs/IBE250/World-is-flat-excerpts-Altermatt-on-Friedman.pdf. Accessed September 5, 2014.
25. Gencer, Ebru A. (2013). "The Impact of Globalization on Disaster Risk Trends: A Macro and Urban Scale Analysis." United Nations Office for Disaster Risk Reduction. http://www.preventionweb.net/english/hyogo/gar/2013/en/bgdocs/Gencer,%202012.pdf. Accessed September 5, 2014.
26. Munro, Emily. (2006). "Natural Disasters, Globalization, and the Implications for Global Security." Geneva Centre for Security Policy. http://www.academia.edu/3187135/Natural_Disasters_Globalization_and_the_Implications_for_Global_Security. Accessed September 6, 2014.
27. Wilcox, Laird. (n.d.). "Laird Wilcox on Extremist Traits." Laird Wilcox. http://www.lairdwilcox.com/news/hoaxerproject.html. Accessed September 8, 2014.
28. Wilcox, Laird. (1987). "What Is Political 'Extremism'?" *The Voluntaryist.* http://voluntaryist.com/articles/027a.html#.VA5MNfmwJcS. Accessed September 8, 2014.
29. Lake, David A. (n.d.). "Rational Extremism: Understanding Terrorism in the Twenty-first Century." David A. Lake Publications. https://quote.ucsd.edu/lake/files/2014/06/Rational-Extremism.pdf. Accessed September 8, 2014.
30. Juergensmeyer, Mark. (2013). *Terror in the Mind of God: The Global Rise of Terrorism.* Los Angeles: University of California Press.

31. Moore, John. (2014). "The Rise of Islamic Terrorism: An Overview." PBS. http://www.pbs.org/wgbh/pages/frontline/shows/target/etc/modern.html. Accessed September 9, 2014.
32. Williamson, Nate. (2014). "American Imperialism and the Rise of Islamic Extremism in Iraq and Syria." Center for Research on Globalization. http://www.globalresearch.ca/american-imperialism-and-the-rise-of-islamic-extremism-in-syria-and-iraq/5400508?utm_source=rss&utm_medium=rss&utm_campaign=american-imperialism-and-the-rise-of-islamic-extremism-in-syria-and-iraq. Accessed September 9, 2014.
33. Fulford, Robert. (2014). "Meet ISIS, the Most Dangerous Terrorist Group in the World." *National Post.* http://fullcomment.nationalpost.com/2014/09/06/robert-fulford-meet-isis-the-most-dangerous-terrorist-group-in-the-world/. Accessed September 9, 2014.
34. Hafiz, Yasmine. (2014). "Muslims Worldwide Fear the Rise of Islamic Extremism: Pew Survey." *Huffington Post.* http://www.huffingtonpost.com/2014/07/02/muslims-against-extremism-pew-survey_n_5551693.html. Accessed September 9, 2014.
35. "Non-Muslims Carried Out More Than 90% of All Terrorist Attacks in the United States." Washington Blog. http://www.washingtonsblog.com/2013/05/muslims-only-carried-out-2-5-percent-of-terrorist-attacks-on-u-s-soil-between-1970-and-2012.html. Accessed September 9, 2014.
36. Henderson, Al. (2013). "10 Worst Examples of Christian or Far-Right Terrorism." *Salon*. http://www.salon.com/2013/08/03/the_10_worst_examples_of_christian_or_far_right_terrorism_partner/. Accessed September 10, 2014.
37. Husain, Rahat. (2014). "In Tragic Twist, Anti-Balaka Christian Terror Groups Attack African Muslims." *Communities Digital News.* http://www.commdiginews.com/world-news/in-tragic-twist-anti-balaka-christian-terror-groups-attack-african-muslims-9691/. Accessed September 10, 2014.

13

Cybersecurity and Protection

Connectivity creates ubiquity…

~Adam Stone [1]

Sci-Fi movies have warned us again and again; sooner or later, our technology will destroy us.

~Adam Stone [2]

CYBER

"Cyber" is a simple prefix often associated with anything related to computers or the information exchanged between computer systems. However, the history and original meaning of the term "cyber" is much more interesting, complex, and ultimately recent; specifically, the "cyber" prefix derived from the term "cybernetics," which was an obscure term popularized by a mathematician named Norbert Weiner in the 1940s. Specifically, Weiner borrowed the ancient Greek word "cyber" that related to the idea of government or governing. Prior to Weiner's usage of cybernetics, the Greek word was always used in writings related to political theory or the science of governance. Instead of following the linguistic origins, Weiner described (what was at the time) a futuristic view that one day there would be a computer system that self-governed its operations [3]. The cybernetics term remained underground until the 1980s and 1990s when music, books, and movies created a new (and still rebellious) movement around the still futuristic approach to the integration of humans and computers that quickly became known as "cyberpunk" [3].

Additionally, a lexicographer with the Oxford English Dictionary recently conducted a thorough review of the history of the term "cyber." This study found that the term underwent significant diversification during the mid to late 1990s. This diversification correlates with the explosion of the World Wide Web and the earliest foundations of the Internet. Consequently, it was during this period that words like cyberbully, cybercommunity, cybergeek, cyberstalker, cybersex, and cyberwar came into existence [3]. Similarly, the term cybersecurity was first coined in 1989 [3]. However, the concept of cyberterrorism did not come into common usage or understanding until the establishment of worldwide terrorism after the September 11 attacks. It is this last component that will be expanded upon throughout this chapter.

By the time of the writing of this book, the concept and number of occurrences of cyberterrorism and the need for cybersecurity have become much more widespread. These occurrences have impacted a wide range of individuals and entities including major global corporations and various governmental components around the world. As a formal terminology, cyberterrorism was first utilized (although not widely) in the 1980s to discuss the dynamic of terrorism (not yet widely understood) of physical impacts to virtual realms. Since then, groups like the Center for Strategic and International Studies (CSIS) have defined cyberterrorism as "the use of computer network tools to shut down critical national infrastructures" such as energy, transportation, and government operations to "coerce or intimidate a government or civilian population" [4]. Another variation of this definition is the "intimidation of civilian enterprise through the use of high technology to bring about political, religious, or ideological aims...that result in disabling or deleting critical infrastructure data or information" [4]. Either way, cyberterrorism—like any form of terrorism—is utilizing an available tool to create havoc and mayhem to areas or people of disagreement.

According to the US Government Accountability Office (GAO), the number of significant cybersecurity events against the US government has increased by 680% since 2006 [1] (see Figure 13.1). For example, in 2012 there were two major cyberattacks against state or federal government departments. The first was a cyberattack against the South Carolina Department of Revenue, which ultimately exposed 4.2 million social security numbers and cost the state nearly $14 million in restoration processes and resulted in the resignation of the agency director. The second major cyberattack compromised the Medicaid data of more than 800,000 residents of the state of Utah and led to the departure of Utah's chief

Figure 13.1 According to the US General Accountability Office (GAO), the number of significant cybersecurity events against the US government has increased by 680% since 2006. (Source: US Department of Defense.)

information officer (CIO) [5]. These sorts of attacks also impact businesses and schools. For example, the University of Nebraska was attacked, which led to the exposure of personal information of approximately 650,000 students, parents, and alumni (see Figure 13.2). In all, one study by the National Cyber Security Alliance reported that the US government unintentionally exposed more than 94 million records containing personal information from 2012 to 2014 [5].

If this is not concerning enough for the emergency management and homeland security industry, it is clear from recent historical examples that the emergency notification and warning systems considered so critical in many communities are also vulnerable to cyberattacks. For example, in 1996, a 19-year-old Swedish man hacked into the telephone network using a computer connection and was able to generate multiple, simultaneous calls (as many as 60,000 total) to the 9-1-1 systems in 11 west-central Florida counties. By making the simultaneous calls into the dispatch center, he was able to tie up all lines available for real and legitimate calls [6]. More recently, a 2013 cyberterrorism attack on the phone system in Spartanburg,

Figure 13.2 The University of Nebraska was attacked by a cyberagent which exposed personal information for nearly 650,000 students, parents, and alumni. (University of Nebraska, Lincoln.)

South Carolina, flooded nonemergency government phone lines, which caused the calls to roll over to the 9-1-1 system and again interfere with the dispatchers' ability to respond to actual crisis calls [1] (see Figure 13.3). Likewise, in 2009, an anonymous hacker changed public safety messages on two LED road signs in Austin, Texas, to warnings of zombie attacks including "The End is Near!!!" "Zombies in Area—Run!" and "Nazi Zombies, Run!" [7]. While the faux-zombie threat is rather lighthearted, it does reinforce the vulnerability of these systems. Lastly, the risks to these systems are not limited to American dispatch centers. For example, in 2013, the website for the Philippines' National Disaster Risk Reduction and Management Council was infiltrated with the website URL being redirected to a pornography site [8].

In Other Words...The Future Threat of Cyberterrorism

President Obama has repeatedly told his aides that there are risks to using—and particularly to overusing...[cyber] weapon[s]. In fact, no country's infrastructure is more dependent on computer systems and thus more vulnerable to attack than that of the United States. It is only a matter of time, most experts believe, before [the United States] becomes the target of the same kind of weapon that the Americans... used secretly against Iran.

~Michael B. Kelley [9]

The use of cyberweapons is not limited to anonymous hackers in remote parts of the world. For example, the Stuxnet virus was identified as the cause of a reported destruction of roughly 20% of Iran's nuclear centrifuges by causing them to spin out of control (see Figure 13.4). Specifically, the virus was inserted into the Iranian system through a worker's thumb drive. In light of the international tensions due to suspicions of nuclear weapon development in Iran, the impacts of this particular event were significant. Moreover, it came out that the Stuxnet virus was actually developed as a joint United States–Israel project [10]. While the effects were significant, the original intent of the virus was not to destroy the centrifuges, but rather reduce the time and long-term

Figure 13.3 Cyberterrorists have attacked local phone systems, which ultimately overwhelmed 9-1-1 emergency dispatching systems. (Source: FEMA/Jason Pack.)

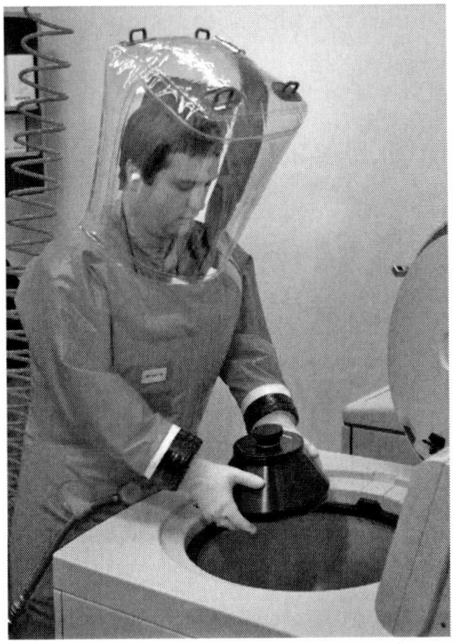

Figure 13.4 The Stuxnet virus was identified as the cause of a reported destruction of roughly 20% of Iran's nuclear centrifuges by causing them to spin out of control. (Source: CDC/Dr. Scott Smith.)

effectiveness of these systems to ultimately limit Iran's development program [10]. This cyberattack was ultimately deemed an "act of force" by research commissioned by NATO's Cooperative Cyber Defense Center of Excellence. Given this designation, this act of cyberterrorism would be illegal according to NATO unless it was done in self-defense [11]. While the United States has never officially acknowledged its role in the Stuxnet attack, it is possible that some would argue such an attack would be qualified as self-defense as well. This type of convoluted understanding only emphasizes the murkiness of cyberterrorism and related protective actions.

VULNERABILITY

Clearly there is risk to computer-based or computer-connected systems. These challenges can be categorized by individual and organizational risks. Likewise, vulnerability can exist on a range of systems including individual computers, master stations, control centers, power connectors, capacitors, voltage regulators, power meters, smart readers (in homes or offices), electrical vehicle charging stations, and many more areas of common and often underprotected systems that contributed to quality of life standards in communities throughout the world. For example, one assessment in 2013 tested 30 different computer-related products from over 20 different vendors and determined that 85% had so-called "low-hanging vulnerabilities" that were either overlooked or underappreciated [2]. On the other end of the spectrum, in 2010 US Secretary of Defense Robert Gates labeled cyberspace (and related vulnerabilities) as the "fifth domain" for potential military conflict along with traditional approaches to land, sea, air, and space protection [12] (see Figure 13.5). If these actions are not clear enough, Norton Security projects that cyber-related threat and crime causes $388 billion in damages globally each year [12].

The widespread interconnectivity of computer systems is a critical part of this consideration. The interconnectivity to critical community support and quality of life is no more evident than in the connectedness of public utilities systems, particularly in developed countries. For example, 26% of public-owned utilities and 28% of investor-owned utilities in the United States are in the planning process of developing an integrated smart grid. Likewise, only 13% of public utilities and 17%

Figure 13.5 In 2010, US Secretary of Defense Robert Gates labeled cyberspace as the "fifth domain" for potential military conflict along with traditional approaches to land, sea, air, and space protection. (Source: US Department of Defense/Cherie A. Thurlby.)

of investor-owned utilities deploy information technology infrastructure updates on a regular basis [2]. According to the utility industry, these public utility systems are particularly vulnerable through the supervisory control and data acquisition (SCADA) protocols which are the communication protocols used to communicate with remote systems and devices. Representatives from the US Department of Homeland Security (DHS) confirmed two such breaches have occurred recently due to system vulnerabilities and brute force attacks on system passwords [13].

In Other Words...Breaches of US Utility Systems

> ...according to...DHS' Industrial Control Systems Cyber Emergency Response Team, or ICS-CERT,..."a public utility was recently compromised when a sophisticated threat actor gained unauthorized access to its control system network.... ICS-CERT validated that the software used to administer the control system assets was accessible via Internet-facing hosts"...connected to a web-connected desktop PC...something that's true of almost every smart grid-enabled control system....
>
> ~Jeff St. John [14]

These types of vulnerability to cyberattacks come in three forms: physical attacks, electronic attacks, and computer network attacks. To this point in the chapter, most of the examples discussed would be categorized as a computer network attack. As loosely established earlier, the first category of network attack is defined as the use of malicious computer code as a weapon to infect computers and exploit weaknesses in software, system configuration, or computer security practices of the individual computer or organizational network. Vulnerability through computer network attacks often allows access to personal or organizational information that otherwise would be restricted [15].

The remaining two types of vulnerabilities are not as common, but must be further considered to understand the full spectrum of cyberterrorism and cyberprotection. For example, the vulnerability from a physical attack would involve the use of conventional weapons to impact, disrupt, or destroy the physical facility maintaining the computers or network components or transmission lines which make connected systems and the Internet possible. Interestingly, only 9% of technology industry experts and researchers surveyed identified physical attacks to infrastructure as a significant concern. This assessment fell well behind malicious network attacks, cloud-computing-based control of resources, and breaches of control organizational information (e.g., misuse of personal information or access) [16].

In Other Words...Unique Physical Risks to Technology Infrastructure

Wide-area fiber-based communication networks pose several unique security challenges. They usually rely on large expanses of minimally to moderately protected fiber infrastructure. They are subject to surreptitious signal monitoring and/or insertion at many points in this infrastructure, which may include optical repeater nodes, switching nodes, operation and management nodes, and fiber itself. They typically carry large volumes of data, making even short duration outages potentially very costly.

~T.H. Shake, B. Hazzard, and D. Marquis [17]

With nearly half of the world's population with access to the Internet, the vulnerability of network infrastructure and transmission lines is increasing year by year. According to Cisco, global Internet traffic increased eightfold from 2005 to 2010 and is projected to quadruple again by 2015 to an astronomical 1 zettabyte of information for the full year (which is roughly equivalent to all movies ever made crossing Internet networks every 4 minutes) [18]. This use has traditionally been heaviest between Europe and North America, but is growing fastest in Africa and portions of Asia. Moreover, the upkeep of these types of transmission lines most commonly comes through the expansion of existing lines rather than replacement of strong, more secure systems due to the economic savings. While this approach can increase the capacity of these lines 25-fold, it can also create significant vulnerabilities due to the patching process [19].

Similarly, local infrastructure systems and power transmission lines are equally taxed and vulnerable to attack. For example, there are more than 5,800 major power generation plants in the United States as well as other smaller power generation stations. To connect these generation stations and move electricity throughout the United States, there are more than 450,000 miles of transmission lines scattered throughout minor and major population centers (see Figure 13.6). Over the next 10 years, the capacity of these lines will grow at least 8%, with increases above 30% in some densely populated areas like Florida and California [20]. Moreover, with nearly $20 billion in annual investment in developing-country energy generation through 2030, the vulnerability of these systems is not limited to the United States [21].

On the other hand, there is a growing diversification of power sources including solar, wind, and geothermal energy. These systems are often highly localized to an individual home or small clustered community. However, there is also a growing trend to allow these systems to be connected back to the power grid and share excess energy generation [22]. Specifically, some green energy and regulatory agencies have estimated that by 2020 solar-power-produced energy will be competitive (without subsidies) with traditional retail electricity prices in a significant portion of the world. If this projected trend holds true, the vulnerability

Figure 13.6 To connect power generation plants, there are more than 450,000 miles of electrical transmission lines throughout the United States. (Source: FEMA/Greg Henshall.)

of electrical transmission systems will no longer be the responsibility of professional energy providers, but will become a shared approach as individuals become both consumers and producers [23].

In addition to the physical attacks to impact computer and infrastructure systems and the direct computer network attacks, the final type of cyberterrorism is from an electronic attack. An electronic attack would use the power of electromagnetic energy (commonly referred to as an electromagnetic pulse [EMP]) to overlap computer circuitry or to insert a stream of malicious digital code through microwave radio transmission. While military, homeland security, and emergency management professionals agree that an electronic attack would be less likely than physical attack or computer network attack, there is still risk from the significant consequences of such an attack. Specifically, history has shown that the detonation of nuclear weapons (both aggressively and in testing) has created these EMP blasts and impacted nearby electronics, circuitry, and computer components. Therefore, since nearly every aspect of life in developed countries has electronic or computerized equipment, the impact of an EMP would be profound. Moreover, given the widespread interconnectivity of transmission systems discussed earlier, there is significant vulnerability from cyberattacks through EMP.

In Other Words...Risk from EMP

EMP does not distinguish between military and civilian systems. Unhardened systems, such as commercial power grids, telecommunications networks, and computing systems, remain vulnerable to widespread outages and upsets.... While the DoD hardens assets it deems vital, no comparable civil program exists. Thus, the detonation of one or a few high-altitude nuclear weapons could result in devastating problems for the entire U.S. commercial infrastructure.

~Dr. George W. Ullrich [24]

COMMON CYBERATTACKS

While physical and electronic attacks on computer- or network-based systems and infrastructure are possible, the computer- or network-based attacks are much more likely. News agencies throughout the world regularly report on large-scale cyberattacks that impact all public sectors. For example, McAfee noted by the end of 2013 that they were detecting and combating more than three new cyberthreats every second with a total of more than 200 million malware varieties in 2013. Unfortunately, this number is twice as many as were detected in 2012 [25]. Likewise, US government officials notified more than 3,000 corporations and businesses that their computer and networking systems had been hacked during 2013 [26]. In the end, these various attacks come in many forms including denial of service (DoS), socially engineered trojans, unpatched software, phishing schemes, network worms, viruses, critical infrastructure attacks, brute force password attacks, and advanced persistent threats. While each of these is unique and capitalizes on various vulnerabilities, some of the more pressing and insidious will be considered further.

The first computer or network attack considered in this section will be a DoS. A DoS attack is a malicious attempt to deny user access to a server or network, thus temporarily interrupting the services offered via the company hosting the network. Typically, this process involves the flooding of the targeted server with external communication requests which overloads the network and limits its ability to respond to legitimate traffic. For example, if Amazon (or a similar online retailer) servers were hit by a DoS attack, the servers would become overwhelmed by the false traffic and become unable (or minimally slowed) to respond to real traffic to the website. DoS attacks are low cost and often difficult to counter without the specific tools. DoS attacks often originate from people with a personal grudge against the host company, competitors looking for a market advantage, or criminals seeking to extort money. Consequently, systems connected to external networks (e.g., Internet) are vulnerable [27]. This includes public safety components like 9-1-1 systems that overwhelmingly maintain closed-loop setups, but still must connect to external systems to ensure efficiency and redundancy.

Unfortunately, research has indicated that the rate of DoS attacks is increasing. Specifically, the volume of attacks from 2012 to 2013 rose by more than 30% for the year and on a month-by-month comparison. Moreover, the complexity of these attacks during that same period of time increased by as much as 42%. Additionally, mobile devices and related applications began to be leveraged as part of the DoS attack [28]. It is clear from these statistics and anecdotal public examples that DoS attacks are increasing across the world and that few technical mitigation strategies work effectively all the time and are available to organizations regardless of size or type. Moreover, there are even fewer nontechnical mitigation strategies that can be deployed by emergency managers not working with terminal government agencies (e.g., the Federal Emergency Management Agency).

In addition to DoS attacks, phishing scams are also becoming increasingly impactful. Phishing is a method for thieves to steal personal information from individuals by pretending to be a company, organization, person, or other entity that the victim would know and trust. The name is derived from the concept of fishing, which lures a fish to the hook by using bait that looks like the real thing. In phishing scenarios, the represented agencies (or "bait") might include family, churches, employers, insurance companies, banks, etc. Consequently, the individual often shares important and otherwise secured information with those individuals acting like a trusted agent. This information is then utilized to access bank accounts, organizational networks, or other secured areas of information. Security experts estimate that US businesses lose approximately $2 billion per year from clients becoming phishing victims [29]. Phishing scams are very common and often target widespread networks with many users such as a school, government, or business.

While there are often tell-tale signs that messages are phishing rather than genuine (misspelled words, poor language syntax and grammar, unreasonable requests, etc.), these types of efforts become increasingly complex even as people are educated about their risk. For example, in 2011, Symantec blocked 1 in every 300 e-mails as a phishing attack with more than 85% presenting themselves as a financial institution. Unfortunately, these phishing attacks have also occurred during major emergencies and disasters like the 2011 Japanese earthquake and related tsunami, where criminals distributed fake messages spoofed as legitimate disaster relief charities to "exploit people's sympathies for the victims of disasters" [30]. Emergency managers and homeland security professionals should assume this type of fraudulent behavior will occur after any disaster no matter the size or impact.

The last form of computer or network attack is malware. This is an all-encompassing term for any software that gets installed on a computer (of any function) that performs unwanted tasks, such as benefiting a third party. These types of programs can range from insignificant inconveniences (e.g., pop-up advertisements) to full-blown computer invasions leading to data or network breaches. Malware can invade a computer if bundled with other software (aka Trojan horse), misleading technology programs (e.g., Internet speed measurements), e-mail links, and through Internet browser security holes [31] (see Figure 13.7). These types

Figure 13.7 The ancient legend of the Trojan horse serves as an excellent analogy for some of the most common types of malware. (National Gallery.)

of software are all malicious and when used individually or collectively allow perpetrators to gain tremendous access to user information.

Specific types of malware programming include viruses, adware, spyware, and browser hijacking software. Viruses replicate themselves to other computers or are programmed to damage computer files, reformat disk space, or use computer memory. Adware is financially supported advertisements that show when connected to the Internet. Spyware gathers information (e.g., websites visited and system information) without the knowledge of the user and transmits it to interested parties. Lastly, the browser hijacking software is adware that changes browser settings (e.g., default home page or search bars) and collects web usage information or directs users to certain websites or functions [31]. These classifications and definitions merely superficially explain and categorize malware, which can be divided into more than 35 different categories based on functionality, purpose, and impact potential [32].

To put the threat and impact of malware attacks in perspective, it is critical to note some of the major incidents in recent years. For example, in 2014, major home retailer Home Depot was hit by malware that affected their point of sale (POS) system, which resulted in the exposure of 56 million consumer credit card accounts. This particular breach came on the heels of a similar breach at merchandiser Target, where 40 million consumer credit cards were exposed [33]. Moreover, major Internet security observers like Kaspersky see nearly 500,000 different forms of malware per year between computers and mobile computer devices [34]. While the Target and Home Depot events are extreme examples, clearly the presence of unknown malware can result in serious security breaches of not only customer data (e.g., credits cards and personal identification numbers), but also critical planning or infrastructure as well. This will be a critical area for emergency managers to monitor as these types of cyber-threats clearly create extreme vulnerability and, unfortunately, are shared across a diversified spectrum of organizations or businesses that may be isolated for practical and commercial reasons.

CYBERPROTECTION

The cyberrisks and -threats discussed throughout the chapter ultimately are perpetrated by individuals using cybertools or equipment to create havoc in computer-based systems. These individuals are colloquially called hackers and ultimately have a variety of motivations. In nearly all cases these individuals attempt to be and often remain fully anonymous, which allows those motivations to include criminal actions, power trips, and in some cases political maneuvering. Because of these ranges of motivations, these individuals can simply work individually or as part of larger, more organized groups.

For example, one of the most well-known hacker groups refers to themselves as Anonymous. Anonymous is a so-called "hacktivist" group that is highly informal and decentralized [35]. To emphasize their rebellious and watchdog mentality, Anonymous graphics and public members often wear Guy Fawkes masks (traditionally the face of the so-called 1605 Gunpowder Plot, which attempted to assassinate England's King James I) [36]. They often target government or corporations that they believe are corrupt (see Figure 13.8). Early Anonymous attacks included the Church of Scientology and the countries of Iran and Australia [35]. Since that time, numerous other entities have either directly or indirectly been impacted by real or threatened cyberattacks by Anonymous. In general, "Anonymous has long demonstrated a transgressive, goading, and impertinent streak...

Figure 13.8 The Anonymous hacktivist group utilizes Guy Fawkes masks to represent their cause of attacking organizations and government agencies they deem corrupt. (Source: James Harrison.)

[with a] freewheeling rebellious spirit" that is ultimately attractive to socially conscious individuals with significant computer skills [37].

In Other Words...What Is a Hacktivist?

Hacktivist is a portmanteau of "hacker" and "activist." When people have technical skills, have access to the Internet and understand how network infrastructure and servers work, it can be tempting to put that knowledge into having some effect on the world. The "activist" part of "hacktivist" means that they don't do their hacking and cracking without a cause. The various people...are united in a belief that corporations and organizations they consider corrupt should be attacked.

~Kim Crawley [35]

Unfortunately, not all Anonymous (or other hacktivist) activities result in positive changes. When hacktivist groups hack government or business records, they often expose personal information of customers, employees, and other related individuals, which in turn is sold in virtual "black markets." For example, Hold Security found that a Russian crime ring had amassed 1.2 billion stolen usernames and password combinations. Likewise, US federal prosecutors charged a Vietnamese identify theft service with stealing as many as 200 million personal records including social security numbers, credits card data, and bank account information. In this case, the Vietnamese company also held more than 540 million e-mail addresses, which were used in association with the personal data for various nefarious purposes [38]. Furthermore, for perspective, Chinese authorities found several virtual locations where as many as 50,000 iTunes accounts (and related personal information) were being sold for up to $10 each [39].

In contrast to Anonymous, some cyberattacks are more formalized, yet still often perpetrated by unofficial or anonymous groups. As the tools for cybermanagement, cyberprotection, and cyberattack become increasingly impactful, some large businesses and most governments in the developed world maintain similar capabilities. For example, the British Government Communications Headquarters (GCHQ) (which serves as a British counterpart to the US National Security Agency) launched a DoS attack on Anonymous hacktivists during a recent campaign, which represented the first Western government known to have conducted such an attack [40]. These DoS attacks were deemed to be "active covert Internet operations" and "covert technical operations" and ultimately led to the jamming of phones, computers, and e-mail accounts through the masquerading of government hackers under a "false flag" operation [40].

Likewise, when Russia invaded Crimea in western Ukraine in 2014, the Russian government utilized targeted insertion of malware, DoS attacks, and various puppet online users to spread disinformation through social networks. During this particular attack, the impacts were significant, but measured enough not to receive significant attention in the global media. However, previous cyberattacks by the Russian government against Estonia and Georgia were "digital…carpet bombings" that attacked the entire cybergovernment structure through massive online assaults [41]. In all cases, these cyberattacks were significant and synchronized with Russian ground actions and political maneuvers.

Given the growing threat of cyberattacks, there is also a growing focus and dedication to creating cyberprotection systems within businesses and countries. For example, in early 2014, US President

Barack Obama issued an executive order that called to create a cybersecurity framework that enhanced "the security and resilience of the Nation's critical infrastructure and to maintain a cyberenvironment that encourages efficiency, innovation and economic prosperity while promoting safety, security, business confidentiality, privacy and civil liberties" [42]. In the end, this framework was structured by government, but ultimately viewed cyberprotection as a multifaceted approach that transcended government support and included businesses and users alike. This type of approach makes sense, considering the shared and overlapping infrastructure (e.g., Internet) that makes networked systems possible. Consequently, these proposed standards were quickly adopted by states and some businesses and will most likely help define future preparedness for cyberthreats [43].

Although the proposed cybersecurity framework is a positive step toward improving cyberprotection in the future, other technologists have noted that government-led decisions are often insufficient because they "focus on some kind of change in the law to raise regulations and/or lower liabilities" [44]. This approach often ignores the systematic trust barriers that often exist between companies and governments on sharing data related to security. Ironically, within systems that fundamentally increase the free flow and exchange of information, security is not widely shared [44]. To address this component of the cyberprotection gap, one strategist for the New America Foundation suggests creating a government agency similar in format and purpose to the US Centers for Disease Control and Prevention (CDC). For example, the CDC was established in 1946 with a "mission to prevent malaria in the United States" and ultimately succeeded "because it established itself as a hub for research on threats the private market wasn't equipped or motivated to confront" [44]. Consequently, the CDC became a trusted clearinghouse for public and private health providers. This type of approach could be repeated for cyberprotection through research and advocacy as a distinctively different component that those services provided through current intelligence and law enforcement agencies [44].

Like many of the issues discussed in this book, the future looks both challenging and bright for the emergency management and homeland security fields. However, cybersecurity and -protection are no different from some of the other emerging issues that have been discussed. While the issues are real and of global impact, the translation to local issues is often difficult to translate. Smaller jurisdictions with limited resources and long-standing and localized issues will have difficulty with many of these future challenges. However, some of these issues may bring in newly discovered benefits to local communities to better prepare for a variety of issues.

REFERENCES

1. Stone, Adam. (2014). "Cyberattack: The Possibilities Emergency Managers Need to Consider." *Emergency Management Magazine.* http://www.emergencymgmt.com/safety/Cyberattack-Emergency-Managers.html. Accessed September 10, 2014.
2. Stone, Adam. (2014). "Securing the Grid." *Emergency Management Magazine.* 9(4).
3. Newitz, Annalee. (2013). "The Bizarre Evolution of the World 'Cyber.'" io9. http://io9.com/today-cyber-means-war-but-back-in-the-1990s-it-mean-1325671487. Accessed September 10, 2014.
4. Tafoya, William L. (2011). "Cyber Terror." *FBI Law Enforcement Bulletin.* http://www.fbi.gov/stats-services/publications/law-enforcement-bulletin/november-2011/cyber-terror. Accessed September 11, 2014.
5. "New Threats, New Tactics." (2013). The Center for Digital Research. http://images.erepublic.com/documents/CDG13+PCIO+SR_Q3.pdf. Accessed September 12, 2014.
6. "Hacking the 9-1-1 System." (n.d.) *Dispatch Magazine On-Line.* http://www.911dispatch.com/911/history/hacking911.html. Accessed September 12, 2014.
7. Fermose, Jose. (2009). "Austin Road Signs Hacked, Warn of Nazi Zombies and End of World." *Wired.* http://www.wired.com/2009/02/austin-road-sig. Accessed September 12, 2014.
8. Mangosing, Frances. (2013). "NDRRMC Website Hacked at Launch of Disaster Information Project." *Inquirer.* http://technology.inquirer.net/31361/ndrrmc-website-hacked-at-launch-of-disaster-information-project. Accessed September 12, 2014.
9. Kelley, Michael B. (2012). "Obama Administration Admits Cyber Attacks against Iran Are Part of a Joint U.S.–Israel Offensive." *Business Insider.* http://www.businessinsider.com/obama-cyberattacks-us-israeli-against-iran-2012-6#ixzz1wYnaa3jK. Accessed September 13, 2014.
10. Kelley, Michael B. (2013). "The Stuxnet Attack on Iran's Nuclear Plant Was 'Far More Dangerous' Than Previously Thought." *Business Insider.* http://www.businessinsider.com/stuxnet-was-far-more-dangerous-than-previous-thought-2013-11. Accessed September 13, 2014.
11. Zetter, Kim. (2013). "Legal Experts: Stuxnets Attack on Iran Was Illegal 'Act of Force.'" *Wired.* http://www.wired.com/2013/03/stuxnet-act-of-force/. Accessed September 13, 2014.
12. Gore, Al. (2013). *The Future: Six Drivers of Global Change.* New York: Random House.

13. Brocklehurst, Katherine. (2014). "DHS Confirms U.S. Public Utility Control System Was Hacked." The State of Security. http://www.tripwire.com/state-of-security/incident-detection/dhs-confirms-u-s-public-utilitys-control-system-was-hacked/. Accessed September 15, 2014.
14. St. John, Jeff. (2014). "Homeland Security: Utility Control System Was Hacked through 'Brute Force Attack.'" The Energy Collective. http://theenergycollective.com/jeffstjohn/386316/homeland-security-utility-control-system-was-hacked-through-brute-force-attack. Accessed September 15, 2014.
15. "Computer Attack and Cyberterrorism: Vulnerabilities and Policy Issues for Congress." (2005). The Navy Department Library. http://www.history.navy.mil/library/online/computerattack.htm. Accessed September 16, 2014.
16. "Experts Complacent about Network Attacks: Study Shows Physical Attacks to Communications Network Infrastructure Deemed Low Priority Risk." *Science Daily*. http://www.sciencedaily.com/releases/2011/07/110728111452.htm. Accessed September 16, 2014.
17. Shake, T. H., Hazzard, B., and Marquis, D. (n.d.). "Assessing Network Infrastructure Vulnerabilities to Physical Layer Attacks." National Institute of Standards and Technology (NIST). http://csrc.nist.gov/nissc/1999/proceeding/papers/p18.pdf. Accessed September 16, 2014.
18. Brodkin, Jon. (2012). "Bandwith Explosion: As Internet Use Soars, Can Bottlenecks Be Averted?" *Ars Technica*. http://arstechnica.com/business/2012/05/bandwidth-explosion-as-internet-use-soars-can-bottlenecks-be-averted/. Accessed September 18, 2014.
19. "Hooking Up." (2013). *The Economist*. http://www.economist.com/news/international/21571126-new-data-flows-highlight-relative-decline-west-hooking-up. Accessed September 18, 2014.
20. "Failure to Act: The Economic Impact of Current Investment Trends in Electricity Infrastructure." (2011). American Society of Civil Engineers (ASCE). http://www.asce.org/uploadedFiles/Infrastructure/Failure_to_Act/energy_report_FINAL2.pdf. Accessed September 18, 2014.
21. "Private Sector." (2014). USAID. http://www.usaid.gov/powerafrica/partners/private-sector. Accessed September 20, 2014.
22. "How a Grid Connected Solar Panel Works." (2013). *Energy Matters*. http://www.energymatters.com.au/residential-solar/how-solar-power-works. Accessed September 20, 2014.
23. Martin, Chris, Chediak, Mark, and Wells, Ken. (2013). "Why the U.S. Power Grid's Days Are Numbered." *Bloomberg Business Week*. http://www.businessweek.com/articles/2013-08-22/homegrown-green-energy-is-making-power-utilities-irrelevant. Accessed September 20, 2014.
24. St. Sauver, Joe. (2009). "Electromagnetic Pulse." University of Oregon. http://pages.uoregon.edu/joe/infragard-2009/infragard-eugene-2009.pdf. Accessed September 21, 2014.
25. Stevenson, Alastair. (2014). "McAfee Uncovers 200 New Cyber Attacks per Minute in 2013." V3. http://www.v3.co.uk/v3-uk/news/2333178/mcafee-uncovers-200-new-cyber-attacks-per-minute-in-2013. Accessed September 25, 2014.

26. Nakashima, Ellen. (2014). "U.S. Notified 3,000 Companies in 2013 about Cyber Attacks." *Washington Post*. http://www.washingtonpost.com/world/national-security/2014/03/24/74aff686-aed9-11e3-96dc-d6ea14c099f9_story.html. Accessed September 25, 2014.
27. "Denial of Service Attacks." (n.d.). *Incapsula*. http://www.incapsula.com/ddos/ddos-attacks/denial-of-service.html. Accessed September 25, 2014.
28. "Top 10 Denials of Service Attack Trends." (2014). Property Casual 360. http://www.propertycasualty360.com/2014/02/11/top-10-denial-of-service-attack-trends. Accessed September 25, 2014.
29. "Electronic Communication." (2014). LifeLock. http://www.lifelock.com/education/electronic-communication/phishing. Accessed September 25, 2014.
30. "Spam and Fraud Activities." (2014). Symantec. http://www.symantec.com/threatreport/topic.jsp?id=spam_fraud_activity_trends&aid=phishing_activity_trends. Accessed September 25, 2014.
31. "Viruses, Spyware, and Malware." (n.d.). MIT Information Systems and Technology. https://ist.mit.edu/security/malware. Accessed September 25, 2014.
32. "Malware Classifications." (2014). Kaspersky Labs. http://www.kaspersky.com/internet-security-center/threats/malware-classifications. Accessed September 25, 2014.
33. Dean, Matt. (2014). "Home Depot Completes Malware Elimination, Says 56M Cards Were at Risk." *Fox Business*. http://www.foxbusiness.com/technology/2014/09/18/home-depot-completes-malware-elimination-says-56m-cards-were-at-risk/. Accessed September 25, 2014.
34. Funk, Christian and Garnaeva, Maria. (2013). "Kaspersky Security Bulletin 2013. Overall Statistics for 2013." Kaspersky Labs. http://securelist.com/analysis/kaspersky-security-bulletin/58265/kaspersky-security-bulletin-2013-overall-statistics-for-2013/#01. Accessed September 25, 2014.
35. Crawley, Kim. (2011). "A History of Anonymous." InfoSec Institute. http://resources.infosecinstitute.com/a-history-of-anonymous/. Accessed October 2, 2014.
36. Sheets, Connor Adams. (2013). "Anonymous Masks Photos Stirs Controversy on Guy Fawkes Day." *International Business Times*. http://www.ibtimes.com/anonymous-masks-photo-stirs-controversy-guy-fawkes-day-1456790. Accessed October 3, 2014.
37. Parker, Chris. (2014). "Anonymous Unmasked." *Huffington Post*. http://www.huffingtonpost.com/high-times/anonymous-unmasked_b_5065038.html. Accessed October 3, 2014.
38. Perlroth, Nicole. (2014). "Russian Hackers Amass over a Billion Internet Passwords." *New York Times*. http://www.nytimes.com/2014/08/06/technology/russian-gang-said-to-amass-more-than-a-billion-stolen-internet-credentials.html?_r=0. Accessed October 3, 2014.
39. "Hacked iTunes Accounts for Sale Online in China." (2011). Phys.org. http://phys.org/news/2011-01-hacked-itunes-accounts-sale-online.html. Accessed October 3, 2014.

40. Schone, Mark. (2014). "Inside the Mind of Edward Snowden." *NBC News*. http://www.nbcnews.com/feature/edward-snowden-interview/exclusive-snowden-docs-show-uk-spies-attacked-anonymous-hackers-n21361. Accessed October 5, 2014.
41. Carafano, James. (2014). "The Dirty Tricks Putin Is Now Using." *The Daily Signal*. http://dailysignal.com/2014/03/22/the-dirty-tricks-putin-is-now-using/?utm_source=facebook&utm_medium=social. Accessed October 5, 2014.
42. Madhani, Aamer. (2014). "Obama Administration Unveils Cybersecurity Guidelines." *USA Today*. http://www.usatoday.com/story/news/politics/2014/02/12/white-house-cybersecurity-framework-released/5422129. Accessed October 6, 2014.
43. "Governor McAuliffe Announces Virginia Adopts National Cybersecurity Framework." (2014). Governor Terry McAuliffe's Office. https://governor.virginia.gov/newsroom/newsarticle?articleId=3284. Accessed October 6, 2014.
44. Singer, Peter W. (2014). "A CDC for Cybercrime." *Wired*. September 2014.

INDEX

G

I

J

K

T